DESIGN FOR DIVERSITY

STRATEGIES TO SUPPORT SOCIALLY MIXED NEIGHBORHOODS

This book is dedicated to my children

Emma, Lucie, and Thomas

And to Peter Talen (1984–2007), their cousin

DESIGN FOR DIVERSITY

STRATEGIES TO SUPPORT SOCIALLY MIXED NEIGHBORHOODS

Emily Talen

Routledge
Taylor & Francis Group

LONDON AND NEW YORK

Architectural
Press

Architectural Press is an imprint of Routledge
2 Park Square, Milton Park, Abingdon, Oxon, OX14 4RN

711 Third Avenue, New York, NY 10017

Routledge is an imprint of the Taylor & Francis Group, an informa business

First edition 2008

Notice
No responsibility is assumed by the publisher for any injury and/or damage to persons
or property as a matter of products liability, negligence or otherwise, or from any use
or operation of any methods, products, instructions or ideas contained in the material
herein.

British Library Cataloguing in Publication Data
A catalogue record for this book is available from the British Library

Library of Congress Cataloguing in Publication Data
A catalogue record for this book is available from the Library of Congress

ISBN: 978-0-7506-8117-9

For information on all Architectural Press publications
visit our web site at www.routledge.com

Typeset by Charon Tec Ltd (A Macmillan Company), Chennai, India
www.charontec.com

CONTENTS

ACKNOWLEDGMENTS

I would like to gratefully acknowledge the financial support of:

The Graham Foundation for Advanced Studies in the Fine Arts
The Richard H. Driehaus Foundation
The University of Illinois Office of Public Engagement
The Illinois Arts Council

Many thanks also to the friends, students and colleagues who helped me with this project: Sunny Fischer, Charlotte York, Mary Antonakos, Phillip Nyden, Peter Marcuse, Dianne Harris, Roberta Feldman, Michael Conzen, John Norquist, Sandy Sorlien, Laura Hall, DPZ, Michael Mehaffy, Michael Stiehl, Sang Lee, Jay Bieszke, Rebecca Bird, Anu Dasgupta, Sara Egan, Lauren Good, Kim Haire, Brad Lents, Beth McLennan, Eric Halvorsen, David Sidney, Shawna Pedersen, Marina Alvarez, Jaime Allentuck, Sarah Voss, and Kelly Craig. Also many thanks to the people of Blue Island, Berwyn, West Ridge, Portage Park, Irving Park, and Bridgeport, who graciously agreed to give up their time to be interviewed for this project.

A very special thanks to Samantha Singer, for all her help along the way.

Finally, thanks to Luc Anselin, who supports me no matter what direction I take.

1

Diversity and Design

Driving south of Chicago along the historic Lincoln Highway, one drives through the strangest juxtaposition of rich and poor places. Along one stretch of the highway is Ford Heights, considered to be the poorest suburb in Chicago, if not the nation. Residents there had a median household income of $17,500 in 2000. A short drive down the road you come to Park Forest, a middle-class planned community with a median household income that is three times as much. There is no disguising the inequality: at one moment you are passing boarded up buildings and liquor stores; ten minutes later, you are whizzing by Starbucks, Home Depot and packed parking lots in front of strip malls.

One can take almost any road through Chicago and traverse similar, strangely abrupt transitions between where the rich live and where the poor live. What is striking is how visible these differentials are, and how coarse the grain is: one neighborhood is poor, one neighborhood is rich; one White, that one Black, that one Hispanic. This is the common, taken-for-granted, vernacular landscape of America. Regarding where White and Black Americans live, it has been called an 'American Apartheid' (Massey and Denton, 1993).

How, and when, did all this spatial sorting occur? When did it become acceptable for communities along the same stretch of road to have such vastly different social and economic prospects? It was not always so. Historians tell us that socially mixed settlements were the norm until the 19th century, a result of economic necessity: nobles needed to be near their serfs in the medieval city, while workers and owners needed to be close to factories in the early industrial city. People lived where they worked, not where their social class was geographically confined. In many European cities, rich and poor were separated only through vertical zoning within apartment buildings, but out on the street, classes shared the public realm. In Chicago before 1850, 'money bought dry ground, not segregation by income' (Massey and Denton, 1993; Bowden and Kreinberg, 1981, p. 116).

But as industrialization progressed in the 19th century, class consciousness became accentuated. In the American city, where ethnically distinct but economically mixed neighborhoods had been formed, class began to trump ethnic affiliation as the main driver of social geography. Olivier Zunz's (1982) study of Detroit showed how a 'silent social revolution' in the first decades of the 20th century created urban worlds defined more and more by class and industrial production and consumption than by strong ethnic bonding or proximity to workplace. New transportation technologies made the spatial sorting that much easier. There were still ethnic divisions, but 19th century neighborhoods and railroad suburbs formerly composed of multiple classes were being replaced by neighborhoods sorted by class and race.

The ultimate segregation: Cabrini-Green public housing on Chicago's North Side, most of which is being demolished.

Thus the story of 20th century urban America is a story of manifested social division. Industrialization brought rising affluence, the growth of the middle class, cheap cars, cheap oil, highways and government subsidies, which, combined with racial and class intolerance, created a toxic mixture that sparked the most extraordinary sprawling out and spatial sorting of cities the world had ever seen. Separation by use mirrored separation by class, and already by 1900, US cities were exhibiting a segregated land use pattern where commercial and office space occupied the center and residential uses were pushed to the periphery (Jackson, 1985). By the 1920s, the connection between social segregation and cities was set. Harvey Warren Zorbaugh, in his classic 1929 study *The Gold Coast and the Slum*, wrote 'there is no phenomenon more characteristic of city life, as contrasted with the life of the rural community or the village, than that of segregation' (Zorbaugh, 1929, p. 232). Louis Wirth reached a similar conclusion in 'Urbanism as a Way of Life', noting that urban populations were both highly differentiated and increasingly subordinated to mass culture (Wirth, 1938).

We developed whole systems that maximized profits through homogeneity. Global distribution, financial markets and lending institutions structured themselves around homogenous clustering and centralized notions of cultural authority. Clustered social spaces began to dictate business decisions (Metzger, 2000; Weiss, 1988, 2000). And where social institutions made progress in breaking down barriers to integration, the corresponding physical design of cities thwarted the translation of this progress in broader terms. For example, although school desegregation made some headway after 1970, the positive effects were short-lived because the larger society remained segregated outside of the schools (Wells *et al.*, 2004), bolstered by school finance policies that had the effect of spatially segregating households by income (Nechyba, 2004).

It was not that we became worse people. Rather that, prior to the 20th century, social distance was maintained in other, often more perverse ways. Jim Crow laws are a notorious example. But unfortunately – or perhaps inevitably – as one kind of social barrier eroded, another arose in its place. This time, the separations were physical and etched into the built landscape like never before. As Olsen put it, 'It is precisely because today's middle classes are less able to exploit the working classes that our cities have become more segregated than they were' (Olsen, 1986, p. 133). As people were socially freer to move about, the means of separation from each other were encouraged by technological advance, especially the automobile. Spatial mobility and social mobility became one and the same.

Where pre-20th century urban form accommodated social mixing, urban form after the 1920s thwarted the ability of classes to mix, even if they had wanted to. Racial prejudice complicated and accentuated the situation. Middle-class Blacks had few housing opportunities outside of Black neighborhoods since White middle-class neighborhoods were resistant to Blacks moving in, even if they were of the same economic class. In post-World War II America, many Whites believed that the only physical context associated with Blacks was urban blight. That degraded conditions were often a result of crowding due to a lack of alternative housing choices was unlikely to be considered (see Kefalas, 2003; Pattillo-McCoy, 1999). And, since the built environment failed to accommodate increased density gracefully, residents of neighborhoods consisting exclusively of single-family homes became alarmed whenever houses were subdivided to accommodate poor families.

The consequences of failing to deliver an urban framework more supportive of social diversity have been monumental. Most damaging has been the concentration of poor people in the inner city. Now, we can only wonder what might have occurred if we had not retreated from diverse environments that permitted a wider social range, what a commitment to diversity might have meant for those seeking to escape the high real estate costs of the inner city and later, the federal bulldozer.

Suburbs have now become more socially diverse, but they have done so in spite of their physical form, not because of it. What has emerged are suburbs that are 'collectively heterogenous but individually homogenous', where people sort themselves out into 'lifestyle enclaves' (Putnam, 2000). Most suburban neighborhoods are still being built for one social class or another, whereby market segmentation strictly divides neighborhoods into pods of distinct income categories.

Would these segregated social patterns have been different if American settlements had been physically structured to accommodate people of all social and economic means?

Different professions and disciplines will have different ways of addressing the problem of social sorting. My interest – and the subject of this book – is the planning and design response.* I will focus on what I call *place diversity* – the phenomenon of socioeconomically diverse peoples sharing the same neighborhood, where diversity is defined by a mix of

---

*I use the terms 'planning and design' to indicate strategies related to intervening in the form and pattern of the built environment. The strategies relate solely to the physical environment and its alteration. Often the strategies focus on identifying specific places for intervention. This is not 'urban design' in the conventional architect's sense, which is often focused solely on the esthetic experience of a small group of spaces or buildings. My sense of urban design is broader, as will be revealed in this book. The interrelated term 'urbanism' refers to the study, planning and design of streets, spaces and neighborhoods, with an emphasis on urban morphologies, block structure, urban spaces, and their geographic context.

income levels, races, ethnicities, ages, and family types. I will argue that good urban form can help sustain diversity. Just what 'good urban form' means in the context of diversity, and how it is possible to justify such assertions, is the subject of this book.

Place diversity is diversity that exists within the realm of 'everyday life' activities – attending school and shopping for groceries, for example. It concerns neighborhoods, whose pattern, design, and level of resources constitute the 'things that really count' – schools, security, jobs, property values, amenities (Pattillo-McCoy, 1999, p. 30). While I do not use precise spatial definitions for the terms 'neighborhood' and 'community', the idea of place diversity is probably most meaningful at a scale that falls somewhere in between Suttles' (1972) 'defended neighborhood' (a small area possessing a definite identity) and the 'community of limited liability' (a larger, often government-defined district). Neighborhoods are to be distinguished from crowded urban places in downtowns and shopping malls – Millenium Park or The Galleria – that are full of social complexity, but where there is little sense of collective ownership or sharing of space for daily life needs.

This book explores how planning and design could be used to support socially diverse places. To understand the possibilities of a design response, I study places that already *are* socially diverse and suggest

Places like Millenium Park in Chicago have a lot of social diversity, but the encounters are random; there is less sense of collective ownership.

In a neighborhood, social diversity is about collective life and the sharing of space for daily needs, as opposed to random encounters with strangers.

ways that the built environment could be leveraged to support their diverse social makeup. What is the physical context of socially diverse neighborhoods, and what does it mean for the residents who live there? Do diverse places look like Jane Jacobs' Greenwich Village, or do they take on a variety of forms? Have these places been responsive to the needs of a diverse neighborhood, in terms of their physical design? What could be the basis of a shared esthetic in diverse places? Given that place-based identity and affection seem so easily associated with 'landscapes of privilege' and exclusion, how do diverse places find a shared definition of place? (Duncan and Duncan, 2004).

Some planners may be skeptical about the wisdom of using design to support diversity. Partly this is a matter of the divorce between social goals and urban design that has been brewing since the demise of modernist urbanism. Cuthbert may be right: 'major theorists in the discipline present us with concepts of urban form that are unrelated, largely devoid of any social content and alienated from any serious socio-economic and political base' (Cuthbert, 2006, p. 21). Thus Jon Lang, in his recent book *Urban Design: A Typology of Procedures and Products*, states that it is the market that should provide people with choices, and that if the market fails to provide them, well, 'these questions are not urban design ones'. He advocates designing sub-areas 'with one population in mind' as a way to 'avoid conflict', while delegating larger, macro areas as the locus of integration (Lang, 2005, p. 369). Besides the anti-progressiveness involved, the problem with Lang's view is that it leaves diverse places behind, providing them with no specific design support, nor any legitimate claim that an approach tailored to their unique social makeup might be needed.

My basic thesis is that there *are* design principles that can help sustain diverse neighborhoods. Modernist urbanism, with its emphasis on functionalism, automobile accommodation and land use separation, exacerbated the key design requirements of diverse places: mix, connection, and security. As cities expanded and grew in the 20th century, design ideology and technological 'progress' – along with other social and economic factors – fueled a built environment conducive to social separation. The result is that we are now left with a physical framework that seems hostile to, rather than supportive of, social diversity.

Fortunately, some places managed to attract and retain diversity. By focusing on those places, I hope to shed light on the ways that the planning and design of built environments can help them achieve stability. Can we leverage the built environment to promote social diversity? Can we help equalize access to social and economic resources by encouraging an urban form more conducive to an integrated population?

COMPLICATIONS OF THE DIVERSITY IDEAL

American ideals speak directly to the need for place diversity. Not only is social inequality viewed as a significant threat to democratic society (Krugman, 2006; Putnam, 2000), but the idea that it is possible to be 'separate but equal' has been rejected unequivocally. The segregation and separation that has come to characterize the American urban pattern goes against the basic underlying ethos of American idealism – a pluralist society rooted in the notion of human equality. Not only does separation challenge the basic foundation of a democratic society, but it engenders profound differences in *access* – access to financial resources, community services, public facilities, social networks, and political power. In addition, it limits our ability to encourage tolerant attitudes and capitalize on the creative aspects of human diversity.

At the same time, there is a certain acceptance of the inevitability of differentiation and segregation of the kind Park, Burgess and MacKenzie identified more than eighty years ago – proclaiming that 'competition forces associational groupings' (Burgess, 1925, p. 79). They made clear that the result of 'continuous processes of invasions and accommodations' was a subdivided residential pattern of varying classes and associated land values, mores, and degrees of 'civic interest'. Where one neighborhood might be 'conservative, law-abiding, civic-minded', another would be 'vagrant and radical' (Burgess, 1925, pp. 78–79). Such differentiation and segregation developed along racial, linguistic, age, sex and income lines, forming units of communal life that they termed 'natural areas'. Zorbaugh's (1929) study of Chicago's Near Northside showed just how stark the contrast between 'The Gold Coast and the Slum' had become.

Strollers along Devon Avenue in the West Ridge community on the north side of Chicago. The area was used as one of the case studies in this book.

City planners are left trying to balance the inevitability of place homogeneity with the intolerability of segregation. They are buttressed by American platitudes about equality and pluralism, but they are constrained by the assumption that, in American society, it is somehow 'natural' to associate social mobility with spatial mobility. One moves up in the world by moving out to a 'better' neighborhood in a completely different location, a process of assimilation with distinct spatial consequences (Peach, 2001).

Even amongst social critics who are most concerned with the deleterious effects of social segregation and concentrated poverty, it is assumed that what is needed is better spatial mobility, not better planning and design to accommodate diversity. Massey and Denton in their well-known book *American Apartheid* (Massey and Denton, 1993) rightfully lament the persistence of residential segregation among Blacks, but see the inequity as a matter of Blacks being prevented from participating in the 'process of normal spatial mobility'. Neighborhood status conceptualized as moving from 'declining' to 'upgrading' (Taub *et al.*, 1984) legitimizes the idea that neighborhoods can rightly be of one homogenous, socioeconomic type or another.

Planners are in the unfortunate position of having encouraged social and economic sorting in the first place. Lewis Mumford argued half a century ago that the mechanisms largely put into place with the help of planners – zoning and highways, for example – had reduced the city's capacity to foster its primary function of human exchange – 'the maximum interplay of capacities and functions' (Mumford, 1949, p. 38). Modern city planning was devoted to creating an 'armature' of 'conflict avoidance' (Sorkin, 1999, p. 2). Through transportation, land use, zoning, housing, mortgage lending, energy, school finance, and many other types of policies, planners have been notoriously complicit in the sorting of social groups and economic functions. And each system is inter-linked, one feeding into and sustaining the other.

But while planning and design of the built environment are believed to have played a strong role in fostering the patterns of segregation that characterize American cities, design reform is not looked to as a way to reverse the situation. In discussions about how to address the antithesis of place diversity – segregation – American city planning, in its capacity as a profession that plans and designs cities, has been relatively withdrawn. Peter Hall (2002) observed that the problems of inner-city disinvestment, White flight and segregation – the most potent manifestations of non-diversity – are problems that, 'almost unbelievably', city planning has not been called upon to answer. Unlike in other countries, 'Americans are capable of separating problems of social pathology from any discussion of design solutions', focusing instead on 'a bundle of policies' (p. 461) – often only weakly related to city planning. While the concepts of 'place', 'neighborhood structure', and 'spatial pattern' have made their way into the prescriptive debate over what to do about residential segregation and lack of place diversity, the planner's unique contribution to this debate seems to be missing.[1]

Jane Jacobs did the most to articulate the fundamental connection between place and diversity. She pronounced that diversity corresponds to physical forms and patterns that maintain human interactions – relationships and patterns of relationships. Her definition of diversity consisted of a mix of uses, including variety in 'cultural opportunities', the inclusion of a 'variety of scenes', and 'a great variety [in] population and other users' (Jacobs, 1961, pp. 143–151). On the effect of the physical environment on human diversity she was not ambiguous: there are physical qualities that create diversity in uses and users, and this is the basis of a well-functioning, vital and healthy city. Her propositions – mixed primary uses, mixed ages, short blocks, concentration – have been the guiding principles for planners who came of age after modernist urbanism and urban renewal had done their damage.

But in terms of implementation her arguments have been problematic. How can her principles ever be employed in the vast number of places that do not have the kind of concentration she required? After all, Manhattan density is unique in the US. Related to this, how do we reconcile her views connecting diversity and intensity with Lewis Mumford's (1930) observation that 'limitations on size, density, and area are absolutely necessary to effective social intercourse'? How can we explain the fact that social diversity in the US is most often located in places that do not conform to Jacobs' conditions of diversity? How do we deploy Jacobsian conditions for generating diversity in places that could never hope to satisfy her criteria, especially the one about concentration? Clearly, we have to come to terms with the fact that building diversity cannot rest on the core of big cities alone, and that it will involve engaging with the 'semi-suburbs', as Jacobs disdainfully called them. In fact, it is in human settlements outside of large, dense metropolitan cores that place diversity is most encountered.

There are other complications concerning the link between diversity and design. Some believe that diversity is just as likely to reside in one type of environment as another (Bruegeman, 2005). Others view the attempt to connect physical form to social goals as overly controlling (Talen, 1999). And not everyone agrees that place diversity should be a key goal. Highly diverse situations may be regarded as too chaotic, while non-diversity or monocultural conditions may be seen as calming. Then there is the view that if you really want to make people's lives better, your strategy ought to be about changing political and economic structures, not rearranging the form of the built environment. Diversity should be 'grown' in place, not dependent on some form of intra-urban social migration resulting in gentrification and displacement. Jane Jacobs called the former 'unslumming'.[2]

Despite the skepticism, the need to support social mix through the mechanisms of planning and design has long been a concept embedded in city planning idealism. The 20th century began with demands that city reformers do something about the 'monotony' of the slums, and the earliest proposals – those of Ebenezer Howard, for example – called for settlements that were internally focused but complete in their provision of the diverse and essential needs of life. Lewis Mumford, Jane Jacobs, Kevin Lynch, William Whyte, and Eliel Saarinen are only a few of the urbanists who thought deeply about the physical context of diversity.

This tradition continues now. City planners are everywhere rallying against social and economic segregation. They are urging a better understanding of the inequities their policies and practices have caused (see, for example, Thomas and Ritzdorf, 1997; Bollens, 1999; Kraus, 2000; Keating, 2001). They are being admonished to be sensitive to difference (Spain, 1993; Sandercock, 1998; Harwood, 2005), to develop an understanding of the effect of multiculturalism on planning (Burayidi, 2000; Qadeer, 1997), and to better recognize the very different social expectations and customs that emerge when new residents move into and gentrify existing neighborhoods (Freeman, 2005; Meyers, 2007). And those connected to New Urbanism, Smart Growth, Livable Communities and related movements are calling for physical designs that explicitly support diversity.

PLAN OF THE BOOK

This book addresses the physical requirements of socially diverse neighborhoods. There are three main parts. In the first part, I set the stage and make the argument: how separated are we, and how prevalent are diverse neighborhoods? More importantly, why is diversity within one neighborhood important, and what does design have to do with it? The second part of the book lays the groundwork for the design proposals I subsequently present in Part III. Using the City of Chicago and its surrounding suburban areas as a case study, I investigate whether social mixing is related to particular patterns and structures found within the urban environment. I then summarize the results of 85 interviews conducted amongst residents who live or work in six of Chicago's most socially diverse neighborhoods.

In Part III, I present my specific strategies and design proposals. There is much to draw from, including nearly a century of scholarship on the effect of physical form on social/psychological phenomena (e.g., behavior, community formation, social interaction, and sense of place). From the city planner's point of view, the connection between design and social diversity is not whether the built environment creates diversity, but whether diversity thrives better, or can be sustained longer, under certain physical conditions that designers may have some control over. Since we seem to know a great deal about how planning and design directly or indirectly *prevents* diversity – e.g., single use zoning, minimum lot sizes, parking requirements, planned (and gated) communities, privatized amenities, excessive road width – it should not be unreasonable to assume that the opposite effect is possible. This is not about determinism but responsiveness – can the physical environment be designed in such a way that diversity is supported rather than prevented? Are there certain physical contexts that diversity seems to thrive in, relative to others? If Richard Sennett is right, that we have lost our capacity for the diversity of public life and the social disorder it may entail, how might the physical environment be enlisted in the effort to get that capacity back?

My hope is that this book will focus more attention on diverse neighborhoods and their needs. If a stable social and economic diversity represents the pinnacle of urban achievement, such places should be cared for, not left to slowly destabilize and decrease in their level of diversity. By contrast, sprawl, defined by its separation of human realms on all levels, has received an extraordinary amount of attention. It has been measured, poked and prodded in all directions. Wouldn't it make sense to spend at least an equal measure of attention on what could be considered our ultimate planning achievement, the diverse human place?

NOTES

1   Ironically, there has been a proliferation of studies by social theorists about the importance of place, but the solutions are much more about politics, processes and programs than planning and design. The book *Place Matters* (Dreier et al., 2001), for example, whose thesis is that 'where we live makes a big difference in the quality of our lives', sees the problem and solution for American cities as being 'primarily political in nature' (p. xi). This is not an interpretation of *place* that squares with the urbanist's sense of the term.

2   For those advocating the former, there were some hopeful signs in the 1990s. The decline of family poverty in the innercity of Chicago, for example, was found to be strongly associated with an increase in female employment in those areas where poverty concentration decreased, in turn affected by a strong economy, expansion of the Earned Income Tax Credit, and changes in public assistance programs (McDonald, 2004).

PART ONE

THE ARGUMENT

This part sets the stage: how separated are we, and how prevalent are diverse neighborhoods? Where diversity or segregation prevail, what explains these phenomena? Most importantly, why is diversity important, and why is design important as a method for achieving and sustaining it?

I review how the normative ideal of diversity in city planning and design originates from distinct, though interrelated, sources: vitality, economic growth, sustainability and justice. I also chronicle the recurrent difficulties that arise in attempting to develop a normative argument for fostering diversity via city design. As a result, even though the connection between diversity and the physical environment is potentially powerful, the connections have been inadequately developed and the level of effort required underplayed.

2

SEPARATION VS. DIVERSITY

The purpose of this chapter is twofold: first, to present a summary of the current status of social separation and diversity, and second, to review the reasons these patterns exist. Understanding the degree of sorting and separation in the American pattern of settlement is somewhat complicated.

Two recent population changes have been making the most headlines. First is the fact that racial and ethnic diversity in the suburbs has increased (Frey, 2002), and more immigrants now live in the suburbs than in central cities (Singer, 2004). The other event is that many central cities gained population in the 1990s as a result of net in-migration.

Neither of these events necessarily bode well for place diversity. For one thing, the racial and economic gaps between city and suburb, or between one suburb and another, or between one neighborhood and another have widened in the past half-century (Ellen, 2000b; Orfield, 2002). In addition, the increase of population downtown did not necessarily signal the proliferation of new mixed-income and multiracial neighborhoods. Instead, it was often a matter of shifting concentrated poverty from one location to another (Powell, 2002).

CLASS AND INCOME

Separation by class[1] is increasing. Between 1970 and 2000, the proportion of neighborhoods that could be categorized as economically diverse decreased for the 100 largest cities in the US (Galster et al., 2005). In this same period, there was a 30% net increase in class dissimilarity, an increase, especially, in the concentration of affluence at the neighborhood level (Massy and Fischer, 2003). This differed by race; separation between rich and poor increased by 34% among Whites, and by 27% among Blacks.

There is evidence that gentrification does not mix income-levels, but instead creates urban enclaves and magnifies segregation by class (Wyly and Hammel, 2004). Capital investment in the inner city is likely fueling this smaller scale segregation. Generally, investment processes have become more geographically complex, and we now see entirely new forms of gentrification, 'rural', 'suburban', 'new build', and 'super' (Lees, 2003a).

In terms of concentrated poverty, there were increases between 1970 and 1990, and then a leveling off in the 1990s. But declines in class segregation at the macro level, at least for the 60 metropolitan areas studied by Massey and Fischer, were basically offset by increases at the micro-level. The process worked something like this: 'as rich and poor families came to inhabit the same regions, states and metropolitan areas … they simultaneously moved into different neighborhoods' (p. 26). The segregation of the wealthy from the poor was especially pronounced in suburban areas, where 'place-shopping' occurs within suburban rings (Fisher *et al.*, 2004, p. 51).

The gap between rich and poor suburbs increased during the 1990s, although less dramatically than in the 1980s. One main conclusion is that 'economic segregation among municipalities is rising', although this varies depending on the region (Swanstrom *et al.*, 2004). Although the share of poor people living in high-poverty neighborhoods declined in the 1990s, the change occurred only in some regions, and mostly in rural areas and central cities, not the suburbs (Jargowsky, 2003). Thus the progress in the 1990s toward decreasing concentrated poverty in central cities, was met with a concern about the potential increase in poverty in other areas like inner-ring suburbs. Although most high-poverty tracts are still in the central cities, a spatial reorganization has been taking place, such that the share of high-poverty tracts that are suburban increased in the 1990s (Kingsley and Pettit, 2003). Progress in deconcentration is attributed to overall economic health, while 'the underlying development pattern that leads to greater neighborhood stratification' was still very much in evidence in the 1990s (Jargowsky, 2003). As the concentration of poverty shifts from one urban location to another, the poor are still concentrated in poor neighborhoods, and minorities are still likely to be concentrated in the city (Jargowsky, 1997). In suburban Los Angeles, the percentage of poor neighborhoods quadrupled between 1970 and 2000 (McConville and Ong, 2003).

Many point to class-based fears about the lowering of property values as being the main determinant of continued economic segregation (Johnson, 1995). Clark's survey (2001) of residential segregation revealed that separation in cities is more a matter of economic difference than discrimination. Pendall and Caruthers (2003) found that income segregation was related to density, in part due to greater competition for space. But they also found there was more income segregation in metropolitan areas that are larger, have younger housing stock, smaller households, less manufacturing-based industry, and faster growth after 1990. On the other hand, one study found that first-generation immigrants may have higher levels of income-mixing, even though they tend to live in racially segregated enclaves (Alba et al., 2000).

RACE AND ETHNICITY

In a dramatic appraisal of Black segregation, Karl and Alma Taueber's book *Negroes in Cities* (1965) measured the persistence of segregation, despite Black economic progress. Their approach was the dissimilarity measure, which (together with the isolation index) are now the most commonly used measures. Generally speaking, segregation by race, especially Black vs. non-Black, is the most pronounced, followed by class, and then life cycle (Fischer et al., 2004).

On an aggregate level, racial diversity in the suburbs has increased (Frey, 2001; Farley and Frey, 1994). Immergluck and Smith (2003) studied home-buying patterns in Chicago and found that the percentage of Whites buying into all-White neighborhoods had declined in the 1990s. Some scholars are optimistic about the prospects for strengthening neighborhood racial integration, particularly where a neighborhood's 'structural strength' has become a focus rather than a neighborhood's racial makeup (Ellen, 2000a). Black suburbanization has been gradually increasing since the 1920s, and nearly doubled in the past two decades (Wiese, 2004). Although relatively few lived in the suburbs in 1960, 30% of the Black population lived in suburbs by 1999 (Clark, 2001). Between 1970 and 1995, seven million Blacks moved from city to suburb, several million more than had moved from the south to the north in the Great Migration (Freedman, 2004). Minorities now account for 27% of the suburban population, and the number of all White census tracts has declined (Fasenfest et al., 2004). However, change in segregation varies by region and by racial and ethnic category. For Hispanics and Asians, segregation appears to be increasing (Logan et al., 2004). Logan et al. (2001) found that segregation by race and ethnicity increased between 1990 and 2000 for children even though it had decreased for adults.

By 2000, more immigrants lived in the suburbs than in central cities (Singer, 2004), although this varied by region. In the New York area, so-called 'melting pot tracts', where three or more groups account for 20% or more of the population, increased from 64 to 84 between 1990 and 2000 (Mehta, 2003). Still, stable, racially integrated neighborhoods are an exception, and Whites' negative perception of African-Americans continues to thwart neighborhood integration (Meyer, 2000). In addition, the suburbs that Blacks locate to are older, closer, and lower-income (Schneider and Phelan, 1993). Generally, places that have been the most traditionally segregated, or have larger minority populations, are more resistant to integration. Black-White segregation measured in terms of dissimilarity has actually increased in places with large Black populations. Overall, minority neighborhoods are now 'poorer, less safe, and less capable of supporting high-quality public services' (Logan, 2003, p. 254).

Conclusions about segregation levels are very much scale dependent. Massey and Fischer (2003) found that segregation levels decreased at the state and regional levels between 1900 and 1960, but increased at the neighborhood level during this same time period. After 1970, racial segregation leveled off at the state and regional levels, but increased at the neighborhood level for many metropolitan areas. Thus, even if there is a statistical increase in the diversity of income or racial categories spread over a region, at the local level, segregation often increased.

Integration may therefore not exist at a meaningful spatial level. A recent study of 'melting pot suburbs' in the San Francisco Bay Area argued that suburban multiculturalism was mostly a myth, and that suburbs were simply patterning themselves into new forms of segregation (Rehn, 2002). While racial segregation at the level of city vs. suburb has decreased, racial segregation is now reflected at the level of suburb vs. suburb. In short, there is evidence that suburbs are differentiating themselves along race and class lines. Enclaves are being formed, resulting from a simultaneous set of processes in which both segregation and integration are increasing (Clark, 2001). Neither does an increase in aggregate diversity mean an increase in neighborhood integration. In the South, levels of Black suburbanization are relatively high, but they tend to be in the form of clustered housing at the periphery, separate from new White suburbs. And the spatial extent of census tracts may be large enough to include both types of segregated communities, statistically masking the experience of segregation (Rehn, 2002).

OTHER FORMS OF SEGREGATION

Schools continue to be a significant source of social separation. They are still largely segregated by race and class (Briggs, 2004; Frankenberg et al., 2003). As Americans took stock of progress toward abolishing the 'separate but equal' doctrine of Plessy vs. Ferguson, struck down 50 years ago by the US Supreme Court, they found that 'the Black poor are more hopelessly concentrated in failing urban schools than ever, cut off not only from Whites but from the flourishing Black middle class' (Freedman, 2004). School segregation reflects residential segregation. For the past 15 years, schools have been resegregating, not desegregating, making our nation's children more isolated, not less. This translates to inequality in educational outcome: 'there is a very strong general relationship between segregation by race and poverty and educational inequality on many dimensions' (Frankenberg and Lee, 2003, p. 3). The schools that Black and Hispanic children attend have double the poverty rate of the schools of White children (Logan, 2003).

Separation between people and jobs also continues to be a significant problem. There are wider gaps between where people live and where they work (Ihlanfeldt and Sjoquist, 1998), signifying that John Kain's mismatch hypothesis formulated in the 1960s still applies. Job growth is higher in outlying areas that house fewer poor and fewer minorities. Most new jobs for less-educated, lower-income groups – i.e., jobs in the manufacturing and retailing/wholesaling sectors – are created in the suburbs (Kasarda, 1995). Job seekers dependent on public transportation often have difficulty getting to the places where lower wage jobs are likely to be found.

In many areas, there is also increasing segregation by household type. For example, there has been an increase in the segregation of married from unmarried people, involving a division between suburban vs. center city population (Fischer et al., 2004). Such changes often put additional stress on municipal governments, since the geography of public services may be out of sink with population needs – the locations of schools and health clinics, for example. A recent study of access to services in Chicago, Los Angeles and Washington, DC found that although there is greater access to social services in central cities, the location of providers does not always match the changing demographics of cities appropriately, particularly in census tracts outside of the central core (Allard, 2004). When there is a mismatch of both jobs and services – where there exist 'profound place-based disparities in opportunity structures and social and institutional resources' — there will be a detrimental effect on labor market success (Coulton, 2003, p. 159).

EXPLANATIONS: SEGREGATION

These patterns of segregation, and their causes, have been the focus of much urban scholarship. Separation has been analyzed on the basis of either socioeconomic status, stage in the life cycle, or race/ethnicity (Fischer *et al.*, 2004). Explanations include consumer choice, discrimination in institutions and governance, neighborhood dynamics, and macro-based explanations involving political economy and social change (van Kempen, 2002; Grigsby *et al.*, 1987). Many view the 'divided and polarized' city as a product of state activities (Marcuse and van Kempen, 2002).

Institutions of varying kinds are an essential feature in diverse neighborhoods. This temple and church are located along a main road in the heart of Bridgeport, one of the diverse neighborhoods studied for this book.

Racial segregation has received the most attention, and explanations include individual choice, financial status, and discriminatory practices, with the last category being dominant (Squires *et al.*, 2001). Others have focused on individual behavior and fear (Clark and Dieleman, 1996; Ellin, 1997), or the way in which ethnic and cultural differences bear out in differentiating space and place (Kefalas, 2003). Vale (2000) studied the American tendency to exhibit 'collective ambivalence' when it comes to locating the poor spatially, stemming from their mixed motives of altruism and social control. Some have pointed out that Americans are generally loathe to having neighbors of lower status than themselves, a 'dirty little secret buried in the shelves of social science poverty studies' (Heclo, 1994, p. 422). Fogelson (2005) documented the long history of restrictive covenants put in place to alleviate the 'bourgeois nightmares' of early suburbanites – fear of poor people and racial minorities – while Keating (1988) implicated the workings of suburban governments and developers in fostering segregation.

Economists have argued that social mix is actually a theoretical impossibility, and that 'even with elimination of all institutional practices that hinder spatial integration, market-based factors would still drive some forms of spatial segregation in a metropolitan area' (Wassmer, 2001, p. 2). Spatial segregation is seen as the inevitable result of 'geopolitical fragmentation', but it is also a result of the fact that higher-income households are able to outbid others for locations closest to the most desirable neighborhood attributes, resulting in a clustering of high-income groups (Vandell, 1995). Thus, at a large scale, segregation has been shown to be correlated with jurisdictional boundaries, likely a result of a Tiebout sorting process (Dawkins, 2004). It is also true that since social homogeneity can strengthen social support networks, help protect against discrimination, and help to preserve cultural heritage (Suttles, 1972), these factor into consumer choice.

A wealth of scholarship has focused more specifically on the effect of planning policy and regulation on the isolation of poor and minority groups, notably Anderson's *The Federal Bulldozer* (1964), Frieden and Kaplan's *The Politics of Neglect* (1975), Kushner's *Apartheid in America* (1982), Keating's *The Suburban Racial Dilemma* (1994), and Thomas and Ritzdorf's *Urban Planning and the African-American Community* (1997). As planning historian Larry Gerckens summarized it, 'virtually every American problem, real, imagined, or socio-psychopathic, was "solved" by physical isolation and segregation' (Gerckens, 1994). Zoning codes that allow only single-family housing but prohibit multi-family housing are the most dramatic (and prevalent) but subdivision regulations that require excessive infrastructure can have a similar effect. Less obvious but potentially insidious ways include the designation of 'neighborhood units' (Silver, 1985), public works projects (Caro, 1974), the push for 'cold war utopias' in the form of peripheral, low density development (Mennel, 2004), or even the way in which planning policies fail to support nontraditional family configurations more prevalent amongst African-Americans (Ritzdorf, 1997).

The government role in contributing to concentrated poverty through its spatial biases is well known (Rohe and Freeman, 2001). Incredibly, the Home Owners Loan Corporation (HOLC) and the Federal Housing Authority (FHA) used an underwriting manual that called for investigating whether a neighborhood had a mix of 'incompatible' social and racial groups (Schill and Wachter, 1995). Highway construction financed by the federal government served to isolate and separate neighborhoods, just as the lack of funding for public transportation ensured that the poor would be immobile.

Many blame modernist concepts of urbanism promoted by the *Congres Internationaux d'Architecture Moderne* (International Congresses of Modern Architecture), known as CIAM, as the key instigator of separation (Trancik, 1986). The concepts they promoted are sometimes linked to Tony Garnier and his 1904 *Cite Industrielle*, displayed in Paris, which was

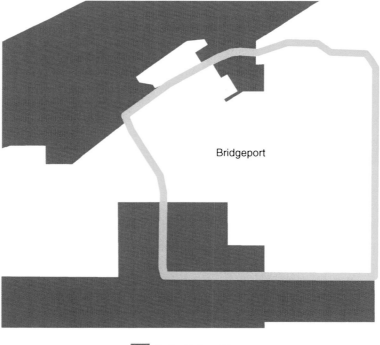

■ Industrial corridor

Bridgeport is surrounded by industrial corridors. This condition probably plays a role in its current status as one of the most diverse places in Cook County.

the first to apply the principles of mass production and industrial efficiency to city form. Garnier's plan boldly rejected past historical styles and offered a 'machine-age community' of hydro-electric plants, aerodomes, and highways, all strictly segregated according to function – building from street and pedestrian from vehicular traffic (LeGates and Stout, 1998, p. xxxi; see also Mumford, 2000). Panerai *et al.* (2004) argued that the gradual demise of the urban block, beginning with Haussmann, was rooted in 'the principle of exclusion' (p. 16).

Postwar suburbanization carried separation by function to an illogical extreme. This condition was nurtured for years by planning organizations, the federal government, and powerful groups like the National Association of Home Builders. In *The Rise of the Community Builders: The American Real Estate Industry and Urban Land Planning* (Weiss, 1987), Weiss showed how community builders helped put in place the deed restrictions, zoning, subdivision regulations, and other land development controls that engendered the segregated pattern of postwar suburbanization. Large tracts of single-family housing often connoted inequity by excluding housing for lower-income groups and failing to provide services that were not automobile-dependent. The shopping malls associated with postwar suburbanization were 'disconnected from the diversity of interests and activities of the 'real world' (Kupfer, 1990, p. 319). The ways in which sprawling cities like Los Angeles have contributed to a segregated urban order are popular themes for social critique, like Mike Davis' *City of Quartz* (1990).

Finally, it must be emphasized that even where segregation is less pronounced, people find other, non-spatial ways of maintaining separation: 'upstairs/downstairs, as in Haussmann's Paris, back of streets and front of streets, as in Engels' Manchester, front building/back building, as in Berlin's Mietskasernen' (Marcuse and van Kempen, 2002, p. 23). Anthropologists have highlighted the various 'codes of deference' that have been used to maintain social separation whenever spatial segregation wasn't practical – rules about clothing, for example (Espino, 2001, p. 1). In the American South, Blacks shared urban space, but the master/servant relationship kept social distances in place. The subordination of Blacks 'paradoxically lessened the need for a rigid system of housing segregation' (Massey and Denton, 1993, pp. 40–41). Gates and walls have replaced mansions, alleys and side streets to carry on the tradition of separating higher status residential areas from lower status areas, even where close spatial proximities remain. Gated communities are a new form of segregation that sort people by age, lifestyle, and other forms of demographic identity (Blakely and Snyder, 1995). Small physical elements can be substantial human barriers too: home security systems, walls, gates, fences and even cyberspace are the 'discourse of urban fear' that encode class separation (Low, 2003, p. 387; Ellin, 1997).

EXPLANATIONS: DIVERSITY

Why, on the other hand, are some places socially diverse? Depending on how diversity is defined, every city has at least some neighborhoods that are diverse, despite the enduring reality that American cities tend to be highly segregated. While there is no explicit definition of the 'socially diverse neighborhood', people consider the mixing of residents by race/ethnicity and by income level or wealth to be the most essential forms, though the mixing of age, family type and household type are also important (Sarkissian, 1976). A diverse neighborhood may have teenagers and elderly; married couples and singles; empty nesters and large families; waiters and teachers as well as professionals; affluent people and people on fixed incomes; and people of varying racial, ethnic and cultural backgrounds. In short, they are places that harbor a full range of human complexity. Rough estimates put the number of neighborhoods that could be characterized as racially/ethnically and economically diverse at anywhere from 5 to 25% of neighborhoods in the US.[2]

Explanations as to why some neighborhoods are socially diverse are likely to be based on three sets of factors: historical/economic/social, policy-related, and physical/locational. Of course, these factors are interrelated. Historical/economic/social factors have an effect on policy, and in turn, some historical/economic/social and policy factors affect physical/locational factors. These interactions are conceptualized shown below.

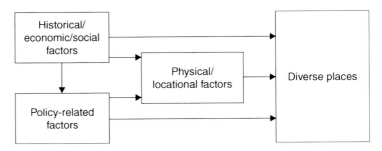

Conceptual framework of factors that explain diversity.

Historical/economic/social factors

Places that are diverse may, first and foremost, be those with a long tradition of diversity. They may be places that functioned historically as immigrant ports of entry, and this openness may have translated into other forms of diversity, such as economic. Because of the historical rootedness of diversity, diverse places may therefore be older than non-diverse places. Older places may be more likely to have experienced a housing filtering process, whereby some proportion of the housing stock, because of its age, became more affordable. At the same time, there may have been an infiltration of new housing stock, creating a mix of building ages conducive to diversity (Jacobs, 1961). Building age mix is also likely to result in a lower median housing age for the area overall, as compared to newer suburban locations. A mix of building ages helps insure a mix of rents and prices (for both owners and renters, and for both residential and non-residential users). Different types of uses require different types, and costs, of buildings.

The dynamics of the local housing market are likely to play a strong role in the effectuation of diversity. It has been shown empirically that racial inclusiveness is related to four housing market conditions: new housing (populated by younger Whites with more tolerance for diversity), multi-family housing, rental housing, and affordable rental housing (Pendall, 2000). These market conditions are also likely to affect income diversity. To the degree that housing unit type is a significant factor in social diversity, diversity is likely to be found where there is a mix of more than one housing type, including owner vs. renter-occupied and single-family vs. multi-family housing.

The employment sector is also likely to be related to residential income diversity. The location of industry seems to have been a particularly important factor. For example, income diverse areas in Chicago and other major cities tend to be primarily in 'blue-collar', ethnic neighborhoods located in inner-ring suburbs (Orfield, 2002). Many of these diverse areas started as industrial suburbs adjacent to railroad lines. Studies have revealed that middle-income suburbs, many of which consist of housing tracts developed near industry, also have the widest range of income groups (Oliver, 2001). Accordingly, the diversity of the older suburb has been heralded as the lynchpin of a new metropolitan structure (Orfield, 2002; Oliver, 2001; Hudnut, 2003).

Such areas may have suffered a certain degree of economic decline in recent decades, particularly due to the loss of industrial jobs (Leigh and Lee, 2004). They may be struggling to recover from the loss of an industrial base, but maintaining some stability either because of the rootedness of the population or because of the influx of new types of activities. Former industrial sites may now function as edges of diverse neighborhoods, in some cases being replaced by luxury town-homes, condominiums, and shopping malls. Income diversity may also result wherever areas with stable land values lie adjacent to areas with decreased land values. This characterizes land near industrial sites and highways, for example.

Diverse areas may be places where macro processes of economic growth did not translate into widespread spatial mobility. Instead, some residents may have been able to 'improve in place'. Or there may have been disruptions in the gentrification process – it may have stalled because of larger economic trends. Although many view the process of middle-class relocation to inner-city neighborhoods as mostly detri-mental (Abu-Lughod, 1994), if gentrification did not result in complete displacement it may have been a significant factor in generating income mix. Some studies have shown this, arguing that gentrification promotes socioeconomic mixing and probably only adversely affects a small number of older residents (Vigdor, 2002; Freeman and Braconi, 2004).

These kinds of processes generate what Nyden *et al.* (1997) refer to as a 'laissez-faire' diverse community. Examples include areas where gentrification stalled because of housing market changes, aging com-munities where residents were replaced with younger people, neigh-borhoods adjacent to fully revitalized areas, places that function as immigrant ports of entry, or the addition of affordable housing develop-ments. Cohen (1998) documented the income diversity of a gentrifying Baltimore neighborhood that resulted when two fractious community groups created an investment standoff.

The social cohesiveness of an area may play a role in promoting or sus-taining diversity. If non-diversity (social separation) is a reflection of the instability of societal interrelationships (Marcuse, 2001), the opposite may also be true: diversity is sustained wherever stable interrelationships occur at a local level. The structural sources for fostering better social mechanisms like 'collective efficacy' are sometimes 'spatially embedded' (Sampson *et al.*, 1999), and the diverse place may be a neighborhood that has been able to capitalize on this.

How far social relations extend in the diverse neighborhood is not clear, although it has been hypothesized that social diversity, not homo-geneity, is the basis of 'active neighboring' (Greenbaum and Greenbaum, 1985). That social relations can be bound by diversity, not similarity, is a theme that some scholars have tried to advance (West, 1991; Sandercock, 1998). Diversity can be seen as generating 'bridging' social capital (Putnam, 2000), a form of social relation that occurs among people of different socioeconomic groups. The power of diversity as a tool for rebuilding social capital has been documented in studies like Warren's *Dry Bones Rattling: Community Building to Revitalize American Democracy* (2001), which chronicled the coming together of rich and poor to work on neighborhood revitalization efforts. At the same time, concentrated disadvantage decreases the capacity of neighbors to work together. Resource deprivation triggers feelings of alienation and dependency, weakening mechanisms of informal social control, a conclusion supported by Velez (2001), who found that disadvantaged neighborhoods most needed to rely on public methods of social control.

Policy-related factors

The deliberate mixing of income groups started in the 19th century as an explicit goal of social reformers concerned with constructing a socially just city (Talen, 2005). New town development in the form of Garden Cities was intended to mix people of various backgrounds by integrating different types of housing units within the same block. Design strategies focused on minimizing the outward variation of housing type; for example, by making apartment dwellings look like large single-family homes. Urbanists continue to argue that there should be no difference in the design and quality of housing for different income categories (Brophy and Smith, 1997).

Policies that are designed to inhibit gentrification and displacement may have the effect of encouraging social diversity. These may include rent control and strategies for subsidizing low-income housing (Kennedy and Leonard, 2001). Or there may be efforts to keep gentrification confined to small areas. Powell (1999, p. 12) asserts that 'revitalization strategies relying on infill or partial/small-scale gentrification' may have the potential to create stable, mixed-income neighborhoods. Policies that seek to increase housing in downtown areas may have a similar effect. These include adaptive reuse, infill on reclaimed land, or niche housing for seniors or students (Birch, 2002). It has to be recognized, however, that infill policies which focus on increasing density could have the effect of increasing social segregation (Huie and Frisbie, 2000). Infill development does not always address affordability problems, as one quantitative study of housing patterns across the US showed (Steinacker, 2003).

A number of recent studies have focused on pro-active efforts at income mixing undertaken through government programs like the

Department of Housing and Urban Development's HOPE VI and Section 8 housing programs (Schwartz and Tajbakhsh, 1997). Social diversity is promoted by either constructing new mixed-income housing developments (the HOPE VI program); by dispersing public housing in higher-income neighborhoods through scattered-site housing; or by giving section 8 vouchers so that tenants can rent housing in mixed-income neighborhoods. Subsidized housing is therefore to be dispersed by demolishing public housing, relying more on tenant vouchers, allowing a greater income-mix in new subsidized developments, or dispersing the spatial location of new public housing projects (Goetz, 1996). The latter category characterizes Chicago's Gautreaux program, under which 75% of public housing families were moved to the suburbs (Squires et al., 2001). HUD's Moving To Opportunity (MTO) program assisted 4,610 families in five cities to move out of high-poverty neighborhoods to more economically diverse ones.

Empirical studies of these publicly-funded attempts to mix income levels have been generally positive, although there are complaints that the projects do not always go far enough (Popkin et al., 2004), and that mixed-income housing should not be seen as a 'silver bullet' that overcomes poverty (Smith, 2002). A study of the Lake Parc Place mixed-income housing development in Chicago found that the development 'accomplished the prerequisites for making mixed-income housing into a community' (Rosenbaum et al., 1998, p. 703). A study of 'scattered-site' public housing showed that residents of target neighborhoods did not 'flee' or engage in panic selling (Briggs et al., 1999), and Feins and Shroder (2005) showed that the MTO program resulted in positive improvements and no loss in social ties for residents relocated from poor to non-poor areas. Another study showed that the deconcentration of assisted housing can have 'positive or insignificant effects' in terms of property value and crime, if the target neighborhood is not already low-valued (Galster et al., 2003).

Studies have been made of pro-active efforts to maintain racially integrated neighborhoods, which have been referred to as 'diversity-by-direction communities' (Peterman and Nyden, 2001). Such deliberateness tends to focus on Black-White integration. Planned diversity can either be a matter of government policy (e.g., Oak Park, Illinois' 'Diversity Assurance Program') or a result of grass-roots effort (the diversity of Park Hill, Colorado, was fostered by an organization of church leaders).

Keating (1994) has written about the implementation and feasibility of housing diversity policies in the Cleveland area. Juliet Saltman

(1990) studied the neighborhood stabilization efforts of five locales and found that the 'fragile movement' was buoyed by progressive non-profit organizations, school boards, and neighborhood groups, working in concert with pro-integrative governments. Smith (1993) and Yinger (1995) reviewed the policies of pro-active integration efforts and found strategies to improve information flow (counseling, for example), pro-active marketing, and more accurate information about neighborhood racial composition. Market intermediaries or housing counselors have been enlisted to act as 'brokering organizations' for cultivating integration (Briggs, 2001, p. 58). Other strategies for stabilizing diversity have included code enforcement, antiblockbusting ordinances, bans on for sale signs, or even grants to individuals who work to support integration.

Regionalism, public transportation, and 'open and democratic local governing' are also believed to be ways of 'integrating diverse groups and cultural practices in a just and equitable fashion' (Polese and Stren, 2000). Local governments may enact fair share housing programs, requiring new developments to provide a certain number of affordable housing units; for example, in Montgomery County, Maryland, 15% of housing units in new subdivisions must be moderately priced, i.e., affordable to working-class households. Once income (or racial) mixing is in place, local governments may require that strategies be enacted to ensure that mixed-income housing projects are supported, through tenant screening, counseling, and project management.

Finally, communities can adopt zoning ordinances intended to encourage diversity. A key component might be mixed housing unit type. Growth controls (building moratoria, permit caps and quotas) have the effect of preventing income and racial integration (Pendall, 2000; Downs, 2000), so changing these rules is believed to have a positive effect on diversity. It is a matter of reversing the rules by which social segregation occurred: allowing multi-family units where they have been excluded, and eliminating rigid building codes, minimum lot size, maximum density, minimum setbacks, and other barriers to infill development. This may also require changing lending policy, capital improvements budgeting, and the protocols of the home-building industry, which tend to favor large, single-use and single-type developments.

Physical/locational factors

Certain physical/locational forms and patterns are believed to be associated with social diversity. Many of these characteristics can be seen as outcomes of the historical/economic/social and policy-related factors already discussed.

Exploring the link between the built environment and social diversity has been revealed in rich descriptive geographic and enthnographic studies. Geographers have been interested in the patterns of social diversity emerging in cities, revealing a complex urban landscape (Bourne and Ley, 2002). The built environment reinforces cultural stereotypes and fears about diversity, at the same time that it provides an outlet for symbolic expression and the creation of ethnic identity (Hanhorster, 2000).

As already discussed, Jane Jacobs (1961) and Lewis Mumford (1968a) had strong ideas about the physical factors conducive to diversity. Lewis Mumford believed that a healthy diversity required limits on size, density and area. Jacobs stressed the importance of use. Offices, factories, dwellings, and other types of primary uses were essential for bringing people to a place, and secondary uses were essential for serving the people that came. Above all, Jacobs argued, there was a logic to the particular mix of uses that would most likely succeed and produce a healthy urbanism.

While many agree that mixed uses – including public and quasi-public facilities and neighborhood-level commercial enterprises – are essential for sustaining socially mixed communities (Myerson, 2001), finding the appropriate mix to support income diversity can be problematic. As Goetz (1996) cautioned, 'the poor relate to [neighborhood] amenities in ways fundamentally different from more affluent families.' For example, public transportation and affordable daycare are likely to be much more important to poor families (see also Bayer, 2000).

Philip Nyden and colleagues (Nyden et al., 1997; 1998) found a variety of physical factors contributing to 'stable diverse' neighborhoods. Significant factors included whether they had 'attractive physical characteristics', access to public transportation and jobs, land use diversity (stores and restaurants), housing stock variety, proximity to downtown, or the existence of 'social seams' in the form of schools, parks, or a strip of neighborhood stores. Others have stressed the importance of a neighborhood's 'institutional base', particularly religious institutions (Rose, 2000), as a way to promote 'strong cross-status ties in mixed-income neighborhoods' (Clampet-Lundquist, 2004, p. 443).

A variety of housing types in one location is an obvious way that physical form promotes social diversity. Mixing housing unit types can occur in two ways: new, mixed housing type developments, or the infilling of new types of development, either on vacant parcels or through the addition of larger homes or smaller units (over garages, over stores).

In the latter case, forms associated with mixed housing include corner duplexes, walk-up apartments on back streets, smaller lots, and duplexes designed as single-family homes. Putting larger or more expensive housing in lower-income areas through demolition and replacement (so-called 'monster' houses in bungalow neighborhoods), or by restoring housing previously divided into smaller apartments, are development approaches that work in reverse: higher-income housing in lower-income neighborhoods (Lang and Danielson, 2002).

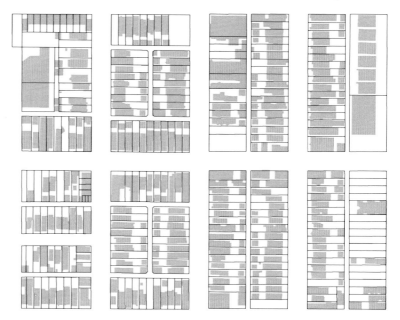

Platting diversity promotes housing type diversity. This example shows parcels and building footprints in the Bridgeport area in Chicago.

A mix of lot types also promotes housing type mix, particularly smaller lots and larger lots in the same block or at least in the same vicinity. Most often, platting in the US, especially post-World War II suburban development, has consisted of strict lot uniformity. Diverse areas in Chicago, however, have a mix of platting arrangements – lot sizes, shapes, and positioning relative to the street – and this has encouraged the mixing of unit sizes and types.

Density also has an effect on diversity. Jacobs preferred densities in the range of 100 dwellings to the acre, and anything significantly lower, she argued, was in danger of producing 'gray areas'. High density and high ground coverage was to be relieved by frequent streets (created by small block size), and variation in building type would have the effect of increasing the diversity of both population and business enterprise. Jacobs argued, further, that this variation would be difficult to achieve wherever land coverage was low and density was high, factors characterizing modern public housing projects.

Despite these calculations, diversity and density do not seem to be correlated in a direct, linear way. As Pendall (2001) explains, density exacerbates segregation by housing type and class because gentrification is more likely to occur in high density neighborhoods where 'proximity-related benefits' increasingly enter 'people's utility functions' (Pendall and Caruthers, 2003, p. 547). As higher-income groups attempt to move 'back to the city', valuing walking and access to amenities (Hughes and Seneca, 2004), enclaves may be formed, as opposed to places with income mix. While some have argued that low density development increases choices for a wider range of socioeconomic groups (Glaeser and Vigdor, 2003), and low density areas have even been shown to be more diverse than the compact city in some cases (Pendall and Carruthers, 2003), low density development may pose a significant problem for low-income people when it comes to the provision of neighborhood-level facilities and access to jobs and urban services.

NOTES

1 'Class' is a term sociologists use to refer to different types of hierarchical distinctions between groups. This can be based on income, wealth, occupation, education, or other variables. Thus class has a broader definition than income. For further reading see Grusky and Szelenyi (2006).
2 Phil Nyden estimated 5–10% at a recent panel discussion on diversity, which took place in Chicago on June 9, 2006 at the I-Space Gallery, 230 W. Superior St. Another view, by Taylor (2006), estimates that 24% of neighborhoods in the Chicago region can be classified as 'diverse'.

3 WHY DIVERSITY?

The notion of 'diversity' defies clear definition. *The New York Times* described it as a trendy, cliché code word (Freedman, 2004), and cultural pundits have tended to complicate rather than clarify the notion. David Brooks (2004) characterized diversity in the exurban fringe as a 'relentlessly aspiring' cultural zone that includes everything from 'lesbian dentists' to 'Iranian McMansions'. The different ways in which diversity is invoked has been described as 'conceptual chaos' fostering a 'diversity of diversities' that is potentially counter-productive (Lees, 2003b, p. 621). For urban planners and designers looking to support neighborhood-level social diversity, less ambiguous definitions are required.

This chapter summarizes the arguments for and against neighborhood-level or place diversity. While this book is based on the view that neighborhood-level social diversity is essential, an understanding of the counter-arguments is necessary. I first summarize the arguments used to support place diversity as a normative goal, organized under two main headings: place vitality and social equity. The categories are inter-related, but nevertheless can be discussed as distinct concepts used to justify not only the need for place diversity, but also the need for a physical environment that can support it.

PLACE VITALITY

There is a pervasive view among urbanists that diversity is a positive force in a global society, a mode of existence that enhances human experience. The city is revered precisely because it is the locus of difference and diversity, and the writings of Fischer (1975), Lefebvre (1991), and Harvey (2000) have put an intellectually rigorous spin on this key point. Even where urban cultural diversity is marketed and sold as a tool of economic development (Lang *et al.*, 1997), the forging of diverse urban lifestyles is nevertheless regarded as an essential asset of cities (Zukin, 1998).

In the realm of urban planning and design, the basic idea is this: 'the combinations of mixtures of activities, not separate uses, are the key to successful urban places' (Montgomery, 1998, p. 98). Thus the notion of 'quality' in the built environment is routinely measured on the basis of variety, choice and interest, all aspects of diversity (Smith *et al.*, 1997). Allan Jacobs and Donald Appleyard (1987) wrote a widely cited manifesto in which they argued that diversity and the integration of activities were necessary parts of 'an urban fabric for an urban life'. Feminists call for building an 'infrastructure for everyday lives', and view separation in space as a reflection of the 'hegemonic assumptions about the respective roles of men and women' (Gilroy and Booth, 1999, p. 307). To some, it is simply that life is 'most rewarding, productive and pleasant' when as many people as possible 'understand, appreciate, and seek out interclass contact and communication in a mode of good will' (Delany, 1999, p. 19).

The maximizing of 'exchange possibilities', both economic and social, is viewed as the key factor of urban quality of life (Greenberg, 1995). More simply, the mixture of housing, schools and shopping is the basic definition of 'a good pedestrian neighborhood' (Hayden, 2003, p. 121). What counted for Jane Jacobs was the 'everyday, ordinary performance in mixing people', forming complex 'pools of use' that would be capable of producing something greater than the sum of their parts (Jacobs, 1961, pp. 164–165). This is similar to the feminist call for an 'infrastructure of everyday life' whereby urban spaces are given multiple roles, not categorized as satisfying either the 'reproductive or productive arena' (Gilroy and Booth, 1999, p. 308).

Lewis Mumford wrote about the importance of social and economic mix often, citing the 'many-sided urban environment' as one with more possibilities for 'the higher forms of human achievement' (1938, p. 486). Planners, in their plans for the physical design of cities, were supposed to foster this wherever possible to achieve the mature city: 'A plan that does not further a daily intermixture of people, classes, activities, works against the best interests of maturity' (Mumford, 1968b, p. 39). By mid-20th century, the rejection of suburbia by planners was based on a perception that it lacked diversity and therefore was 'anathema to intellectual and cultural advance' (Sarkissian, 1976, p. 240).

Mumford's ideas about human diversity were strongly influenced by Patrick Geddes, and both men saw the advantage and positive stimulation of cities as a way to accommodate 'the essential human need for disharmony and conflict' (1938, p. 485). The very existence of diversity signifies democracy in action (Sennett, 1970). As one researcher put it, 'it is discourse over conflict, not unanimity, that helps democracy thrive' (Patricia Gurin, quoted in Skerry, 2002, p. 23). In a similar vein, Baumgartner's (1991) study of the 'moral order of a suburb' revealed the negative implications of a homogenized, privatized social worlds where communal conflict is internalized or avoided rather than dealt with openly.

Diversity is seen as the primary generator of urban vitality because it increases interactions among multiple urban components. A 'close-grained' diversity of uses provides 'constant mutual support', and planning must, Jacobs argued, 'become the science and art of catalyzing and nourishing these close-grained working relationships' (1961, p. 14). Thus the separation of urbanism into components, like land use categories, miles of highways, square footage of office space, park acreage per capita – all of these abstracted calculations lead to, as Mumford termed it, the 'anti-city' (1968b, p. 128). Jacobs similarly berated planners for treating the city as a series of calculations and measurable abstractions that rendered it a problem of 'disorganized complexity', and made planners falsely believe that they could effectively manipulate its individualized parts.

Diversity is usually not thought of as being chaotic or random. For Jane Jacobs, social, economic and physical diversity effectively co-existed within an underlying system of order, which she termed 'organized complexity'. Similarly, Eliel Saarinen (1943, p. 13) thought the diversity of urban elements could be brought into 'a single picture of rhythmic order'. Melvin Webber's 'Order in Diversity' (1953, p. 51–2) essay lamented the mistaking of complexity for chaos. He believed that plans, to accommodate diversity, must be designed 'to accommodate the disparate demands upon land and space made by disparate individuals and groups'.

Most agree that diversity must be substantive, not superficial: 'it is the richness of human variation that gives vitality and color to the human setting' (Raskin, quoted in Jacobs, 1961, p. 229). And in fact, a commercial street that looks garish and chaotic is most likely not diverse but homogenous. Venturi et al. (1977) discovered that architecture will attempt to present a sense of variety by being exhibitionist in the midst of an underlying homogeneity. Extreme variations in color, form and texture are buildings crying out to be recognized amidst an overbearing pattern of sameness.

Diversity is associated with vibrant commercial areas, such as this commercial corridor in the heart of West Ridge, a diverse community on the north side of Chicago.

Economic health

A sub-category of 'place vitality' is economic health. Urban diversity, the 'size, density, and congestion' of cities, was considered by Jacobs to be 'among our most precious economic assets' (Jacobs, 1961, p. 219). There has been disagreement over the role of diversity in generating knowledge spillovers, but the view that diversity of industries in close proximity generates growth, rather than specialization within a given industry, is generally accepted (Glaeser et al., 1992; Quigley, 1998). The richness of human diversity is an economic asset because innovation within firms can come from spillovers outside of the firm. Spillovers depend, to some degree, on spatial proximity, since distance affects knowledge flows (Glaeser, 2000). The scale at which diversity is able to create spillovers that contribute to innovation and vitality depends on the scale at which cross-cultural knowledge spillovers are likely to occur.

Richard Florida has been particularly explicit in arguing for the importance of diversity in economic terms, but his argument is structured differently. His creative capital theory states that high densities of diverse human capital (the proportion of gay households in a region is one measure), not diversity of firms or industries in the conventional economic view, is what promotes innovation and economic growth (Florida, 2002a). Cities that are open to 'diversity of all sorts' are also the ones that 'enjoy higher rates of innovation and high-wage economic growth' (Florida, 2004, p. 1). Cities should therefore attract human capital, focusing on what's good for people rather than on, more conventionally, what's good for business. This naturally leads to an elevation of the qualities of place, since 'talent does not simply show up in a region' (Florida, 2002b, p. 754).

Studies have attempted to show that immigration is an economic stimulus, and that the economic health of the US can be tied to its historical openness to foreigners (Zachary, 2000; Saxenian, 1999). In *The Global Me*, Zachary (2000) proclaims that diversity is the defining characteristic of the wealth of nations, a new world of citizens who possess multiple 'ethnoracial affiliations'. Maignan *et al.* (2003a) point out that it is possible to find a historical relationship between religious tolerance and innovation, starting with the rise of capitalism in the 17th and 18th centuries. Intolerance, on the other hand, correlates with 'crisis and slackness in business' (p. 13). It has also been shown that diversity in metropolitan areas is correlated with lower unemployment and less instability (Malizia and Ke, 1993).

Diversity promotes economic health because it fosters opportunity. In Jacobs' words, cities, if they are diverse, 'offer fertile ground for the plans of thousands of people' (p. 14). Non-diversity offers little hope for future expansion, either in the form of personal growth or economic development. And in fact class segregation has been shown to lower a region's economic growth (Ledebur and Barnes, 1993). Nor are non-diverse places able to support the full range of employment required to sustain a multi-functional human settlement. Diversity of income and education levels means that the people crucial for service employment, including local government workers (police, fire, school teachers), and those employed in the stores and restaurants that cater to a local clientele, should not have to travel from outside the community to be employed there.

Finally, there is the idea that a diverse community is better able to take care of itself. The 'richly differentiated neighborhood' is more 'durable and resilient' against economic downturn. But interaction among diverse peoples also helps generate the contacts needed for individual success. Diversity builds social capital of the bridging kind by widening networks of social interaction. Where there is less social diversity and more segregation, there is likely to be less opportunity for the creation of these wider social networks. This could be a significant disadvantage for segregated neighborhoods, and could even have the effect of prolonging unemployment (Grannoveter, 1983).

Ecology and sustainability

Place vitality may be linked, as well, to ecology. Relating human diversity to biological diversity is a pervasive theme in urban planning, explored first by ecologically-oriented planners like Patrick Geddes (1915) and Lewis Mumford (1925), and later by Ian McHarg (1969). The connections generally revolved around a regionalist approach to city planning. The meaning of an ecologically informed urban planning meant, essentially, a regionalism in which diversity thrived in the form of close-knit communities well-integrated within a larger ecological context. Benton MacKaye (1928) drew creative, if abstract, analogies between the world of planning and the world of nature to make these connections.

Ecology originated as a field of study focused on holistic notions about integrated, balanced, interdependent communities of biological organisms (Calow, 1998). The most famous application of ecology to human environments remains that developed by McKenzie, Park, and other theorists of the Chicago school in the 1920s and 1930s (Park et al., 1925). According to Park, 'all living organisms, plants and animals alike, are bound together in a vast system of interlinked and interdependent lives' (Park, 1952, p. 145). Amos Hawley's (1950) *Human Ecology: A Theory of Community Structure* theorized the community as groups of individuals dependent on each other for survival, and human ecology was essentially the study of the organization of interactions and functional relationships between groups in a community.

The attempt to link the human and natural worlds is often discredited. Sociologists and geographers have questioned the relevance of communities (cities) as functional units, given the global nature of migration and capitalism and the existence of nations, corporations and other entities that operate beyond the bounds of cities (Harvey, 2000; Sassen, 1997; Gottdeiner and Feagin, 1998). Some emphasize the difficulty of the study of community as a unit, since defining the territory of a set of groups as cohesive or independent is problematic, while others criticize human ecology for focusing on human interaction in the abstract, divorced from the physical, spatial environment in which it takes place. Critical theorists like Lefevbre (1991) and Soja (1989) argue that social processes have a fundamentally spatial dimension, occurring in particular places at particular times, which cannot be detached in an abstract manner as in human ecological theory.

Despite these criticisms, the application of ecological theory to human environments continues (Berry, 2001). Recent calls for 'a new paradigm for transdisciplinary landscape ecology' in which 'multifunctionality in landscapes' is the norm is essentially about managing landscapes for multiple uses as a counter-strategy to the Western tendency to segregate functions (Naveh, 2004, p. 33; Brandt and Vejre, 2004). Planners are instructed to 'weave together' a diversity of elements, like 'a quilt held together with threads' in their approach to planning for communities. These principles of landscape ecology then become 'the new civics of sustainability' (Dramstad et al., 1996, p. 5).

Texts on sustainability in cities are likely to feature diversity as a fundamental goal. Beatley and Manning (1997, p. 36) define the sustainable community as 'one in which diversity is tolerated and encouraged', where 'sharp spatial separation or isolation of income and racial groups' is non-existent, and where residents have equal access to 'basic and essential services and facilities'. Steiner posits the notion of 'unity in diversity' as a fundamental principle of human ecology. It is through diversity that pluralist societies, defined as 'heterogenous groups within a space' achieve unity (Steiner, 2002, p. 34). Diversity is what builds resilience in both human societies and natural ecosystems (Capra, 1996).

Characteristics of sustainable cities are almost all tied to, or ultimately derived from, the need for social and economic diversity. The book *Building Sustainable Urban Settlements* (Romaya and Rakodi, 2002), for example, lists 'mixed land uses' first under its set of principles for building sustainable settlements. Reduction of travel costs, and therefore energy consumption, is usually a primary motivation. The 'land use–transport connection' is put forth as a counter-response to the problem of non-diversity, i.e., functional isolation (Newman and Kenworthy, 1996). A mixture of land uses has been shown empirically to encourage non-automobile based modes of travel such as walking and bicycling (Cervero, 1996), which in turn are seen as having a positive impact on public health (Frank et al., 2006).[1]

SOCIAL EQUITY

There are two ways in which social diversity is linked to social equity. First is the idea that social diversity is equitable because it ensures better access to resources for all social groups – it nurtures what is known as the 'geography of opportunity'. In the second sense, diversity is seen as a utopian ideal – that mixing population groups is the ultimate basis of a better, more creative, more tolerant, more peaceful and stable world. Under the first objective, distribution and access to resources is a matter of fairness. Under the second, even those in higher-income brackets can take advantage of the creativity, social capital and cross-fertilization that occurs when people of different backgrounds, income-levels and racial and ethnic groups are mixed. The former speaks to functionality and material need, the latter to the nurturing of the human spirit.

Diverse neighborhoods foster social mixing. Above, a cafe in Portage Park welcomes all ages to participate in an 'open mic night'.

The idea of calculated social mixing in cities and towns – deliberately attempting to put people of different means and backgrounds in the same general area – was born in the 19th century by idealists and social critics who deplored the living conditions of the poor. The Settlement Houses and Co-Partnerships of Samuel and Henrietta Barnett, Octavia

Hill, Jane Addams, and others were aimed at educating and socializing the poor, but also sensitizing the rich through deliberate social mix in urban places. Others focused on constructing utopian communities that deliberately mixed people of different social and economic classes through spatial planning and housing design. Examples include company towns like Bournville or Ebenezer's Garden Cities of Letchworth, Welwyn Garden City, and Wythenshawe. They were meant to include all social groups, although in varying degrees of physical closeness. Whereas Howard's idea of mix was more segregated on a micro-level, Bourneville's mix was fine-grained. But even Howard's idea included a level of mix that far exceeded conventional American patterns: places to work, shop and recreate were to be within a short walking distance. Raymond Unwin, the architect who gave physical form to Howard's theory of Garden Cities, stated that town planning must 'prevent the complete separation of different classes of people' (Unwin, 1920, p. 294). Of course, there were limits: 'close enough, but not too close' as Hall (2002, p. 104) describes Unwin's attempt at social mix.

Sarkissian (1976) reviewed the history of attempting to mix social groups via town planning (see also Cole and Goodchild, 2001). She identified the various goals of social mixing: to raise the standards of the lower classes, to encourage esthetic diversity and cultural cross-fertilization, to increase equality of opportunity, to promote social harmony, to improve the physical functioning of the city (better access to jobs and services), and to maintain stable neighborhoods, whereby one can move up or down in housing expenditure and remain in the same area. Empirical studies of stable, diverse neighborhoods have documented the ability of integration and stability to co-exist (Ellen, 1998; Lee and Wood, 1990; Maly, 2000; Ottensmann, 1995).

Using the physical environment to promote racial justice was an important goal in postwar planning. Books like Charles Abrams' *Forbidden Neighbors* (1955) were about racial justice through integration, arguing the case for residential social mix 'from every conceivable angle' (Sarkissian, 1976, p. 239). Although not heralded as such, urban renewal programs in the fifties were focused on social mix, largely on the grounds that mix could make communities more stable (Glazer, 1959). Housing activist Catherine Bauer held similar views about the importance of social mixing (Bauer, 1951). Following through on these ideals, new towns developed in the 1960s – Columbia, Maryland and Reston, Virginia – were planned for racial and economic mixing.

Sociologists argue that if people are confined to a social environment with concentrated problems, a variety of life chances are diminished, such as access to health care and employment information. Children growing up in neighborhoods of diminished resources are negatively impacted because student achievement is significantly influenced by classmates and their families, not just teachers (Burtless, 1996). Researchers have concluded that the concentration of social disadvantage leads to an increase in crime rates (Jargowsky et al., 2005). Concentrated poverty also correlates with economic dislocation and the loss of jobs, something William Julius Wilson protested in The Truly Disadvantaged (Wilson, 1987).

Because of concentrated social conditions like joblessness, the segregated neighborhood will experience property disinvestment, housing abandonment, and the withdrawal of commercial activity. Kefalas' (2003) study of a working-class neighborhood showed how rascism was being driven by the deterioration of place, where the physical decay of the ghetto was used 'as irrefutable evidence of widespread social breakdown' (p. 52). Loss of consumer income means loss of consumer demand and a depleted retail sector. Rich and poor communities will have different tax structures resulting in different resource levels. High-income residents will seek lower property tax rates, while low-income neighborhoods will have to tax themselves at a much higher rate to receive a comparable level of service (Massey and Fischer, 2003).

One solution to these inequities is to deconcentrate the poor – in other words, promote place diversity. It is argued that poor people who move to low-poverty neighborhoods will benefit by obtaining better quality housing, better resources, more employment opportunities, and access to wider social networks. This is not confined to central cities vs. suburbs; recent revelations that suburban poverty rates are higher than urban poverty rates have produced calls for more economically integrated neighborhoods throughout metropolitan regions (Berube and Kneebone, 2006).

Integration also creates a basis for 'pluralist politics' based on shared benefit (Massey and Denton, 1986, p. 14), whereas non-diverse, exclusively poor neighborhoods compete for public expenditures on their own. Neighborhood-based diversity provides the basis for shared concern, a 'coalition politics based on geographically structured self-interest' (p. 157). Diversity is thus essential for making sure that groups can share interests and build political effectiveness. Social segregation, by limiting this power, limits the degree to which physical improvements – facilities like schools and parks – are likely to be funded. Supporting this are findings that neighborhood public facilities play a role in reducing crime (Peterson et al., 2000), which is important for sustaining mixed-income communities (Myerson, 2001).

The idea that something useful is accomplished by ensuring that divergent populations come together and encounter 'the Other' in collective space may seem contrived. Yet the idea that everyday contact is crucial for 'reconciling and overcoming ethnic cultural differences' (Amin, 2002, p. 959), or that casual social contact or 'weak ties' are essential for community building (Skjaeveland and Garling, 1997) are still compelling ideas. There is also empirical validation. Research has demonstrated the positive effect of desegregated schools, where the mixing of racially diverse groups has been found to significantly lessen fear and distrust of dissimilar people (Wells et al., 2004). Bringing people in contact with one another, sometimes referred to as the 'neighborhood contact hypothesis', is believed to reduce racial prejudice and raise the bar on the neighborhood tipping point (the point at which neighborhoods 'tip' to all one race; see Ihlanfeldt and Scafidi, 2002). Encountering the full spectrum of a community is also thought to be essential for a child's education because it teaches them that they are part of a larger culture, that they have a role in, and can participate in, a shared society. The value of constant encounter was nicely summed up by Michael Sorkin: 'It is no tautology to suggest that the only training for living together is living together' (Sorkin, 1999, p. 7).

COUNTER-ARGUMENTS

To some degree, there is a conflict between the goals of 'vitality' and 'equity', the two main objectives of place diversity. One problem is that the goal of place vitality needs to be matched with a commitment to social equity, but this is not often recognized. Diversity has been embraced as a fundamental concept in the attempt to stimulate an urban renaissance, but liberal notions of cultural pluralism do not sit easily with the business community (Lees, 2003). Many, if not most, mixed-use developments consist of large-scale, high-end retailing and expensive condos in prime locations, often with the assistance of community redevelopment agencies (Culp, 2003). Most often, the public benefit is strictly a matter of stimulating downtown economic growth, not procuring social equity and sustainability through human diversity.

Diversity is a matter of degrees, and it varies in terms of its negative and positive effects. There is no doubt that diversity is a matter of cultural perception, as are density and many other aspects of urban place and form (Rapoport, 1977). In other words, it is not always easy to recognize when diversity is a good thing, and when it is not. In the case of mixing incomes, there may be thresholds that need to be met before the benefits of income mixing have effect. Quercia and Galster (1997) hypothesized that attracting the middle-class to the central city was probably only of benefit if certain levels were reached.

Environmental and social diversity goals may conflict wherever the former is defined as reducing land consumption and increasing density. Increasing density holds a real danger of increasing social segregation (Huie and Frisbie, 2000), and as a result, suburbanization is sometimes viewed as a mediator of segregation (Logan *et al.*, 2004). Jacobs relished the diversity of cities with 'so many people' living 'so close together' in places of 'exuberant variety', but critics thought her glorification of diversity was akin to 'home remedies for urban cancer' (Mumford, 1968). And if diversity depends on a given level of urban concentration, as Jacobs prescribed, it will appeal only to one segment of the population, not to everyone. As higher-income groups attempt to move 'back to the city', valuing walking and access to amenities (Hughes and Seneca, 2004), these kinds of transitions may not be creating diversity but rather shifting enclaves. This underscores the need to understand the difference between 'chosen enclave' and 'enforced ghetto' (Peach, 2001).

Skerry (2002) uncovered a number of studies that have concluded that diversity is a mixed blessing: it poses problems for community policing efforts, it impedes the ability to maintain social cohesion, it correlates with a weak labor movement; and, in general, creates dissension and conflict. He points to a study by psychologists Williams and O'Reilly (1998) that reviewed 40 years of research on racial and ethnic diversity and concluded that diversity is likely to 'impede group functioning'. Research on British towns by Kuper (1953) suggested a correlation between social mix and neighborly disputes. Recent work by Putnam (Sailer, 2007) and Wilson and Taub (2006) have shown significant levels of social stress brought on by living in diverse places. And social mixing can provoke a backlash. The 'relative deprivation' model of neighborhood effects maintains that more affluent neighbors can stimulate resentment among less-affluent neighbors and create the need to maintain a 'deviant subculture' (Mayer and Jencks, 1989, p. 1442).

Separation seems to reflect human nature, and it is often argued that human settlements 'naturally' segregate into homogenous areas (Kostof, 1991). Ethnic segregation may be one of those naturally occurring phenomena. In their classic study of ethnic identity, *Beyond the Melting Pot: the Negroes, Puerto Ricans, Jews, Italians, and Irish of New York City,* Glazer and Moynihan (1963) showed that ethnicity in the American city was not a relic from the era of mass immigration that was rapidly diminishing in importance, but instead was a new phenomenon that arose in response to life in the city. Many believe that multiculturalism and 'ethnic federalism' is essential because it sustains group identity, ethnic heritage, and the support systems that go along with the preservation of cultural affiliation. Assimilation, on the other hand, works on the model of melting pot conformance to White, western values. If a particular concentration of people constitutes a voluntary rather than a coerced enclave, there may be no need for intervention, and in fact the 'voluntary enclave' may be viewed as a legitimate act of empowerment (Peach, 2001). In a similar vein, Vale (1998) argues that income mixing is 'politically and financially appealing but socially unnecessary'. Some have even been arguing against the need for school integration, believing that segregated schools need not be integrated, just equalized (Bell, 2004; Ogletree, 2004).

Social homogeneity can be seen as empowering. In the 1960s, Black militants reacting against forced relocation under urban renewal 'insisted that integration threatened the solidarity of the ghetto, that it was a device to divide and rule' (Sarkissian, 1976, p. 242). Under current public housing programs, a similar reaction is occurring, and the placement of low-income groups in suburban, higher-income locations is seen as having the adverse affect of disrupting the social support systems that developed in public housing projects (Feldman and Stall, 2004). Some argue that Blacks prefer to live in Black neighborhoods where they can establish their own social institutions (Thernstrom and Thernstrom, 1997; Patterson, 1997).

Similarity of population leads to similarity in demand for public goods. Economists see this as efficient Tiebout sorting: a 'lower-level demander' is prevented from 'free riding' off the demands (and payments) of a 'high-level demander' (Wassmer, 2001, p. 15). Ethnic diversity, which ensures heterogeneity of preference, results in lower levels of public goods provision (Alesina et al., 1999). There is also the question of whether the type and spatial location of amenities in high-status areas is appropriate to meet the needs of residents with a different set of requirements. Just putting the poor in an affluent area does not create a better life for them, and may in fact isolate them socially and functionally.

Mixing housing for very different income levels does not always result in a desirable kind of diversity. Pictured above is Cabrini-Green public housing next to new condo development.

Where the promotion of diversity entails gentrification, there are major concerns. The infiltration of higher-income groups may produce an increased income mix, along with an increase in property values and revenues, but it can also mean disruptive changes in local leadership, institutions, social networks, and established cultural anchors (Kennedy and Leonard, 2001). It may mean that low-income residents are pushed to other locations, severely disrupted at best. This has been the view of many urban scholars, who argue that the process of middle-class relocation in inner-city neighborhoods is nothing but loss, the killing of 'neighborhood spirit' (Abu-Lughod, 1994). Grady Clay quotes an eerie description of what can happen if the gentrification process goes full swing: 'Summer chairs on the sidewalk, television out on the stoop, and children's street games are replaced with herringbone pavements, fake gas lamps, wrought iron window railings, and a deathly hush on the street' (Holcomb, 1986, quoted in Clay, 1994, pp. 117–118).

CONCLUSION

I have presented here a wide array of arguments for and against place diversity. Understanding both sides is necessary, and there are indisputable arguments to be made on either side. Sometimes the balance tips strongly in favor of the need to promote place diversity, but there are undoubtedly times when voluntary social segregation may be warranted. Still, whatever advantages there may be to social homogeneity in a neighborhood, they will have to stand up against the problems of isolationism, exclusivity and uneven access to resources. And if there are instances where the goal of diversity can legitimately be downplayed, that will not change the need to support place diversity in many other locations.

Arguments in favor of diversity seem motivated by positive, hopeful views about life in an inclusive society. Arguments in favor of homogeneity seem motivated by doubt, fear, and even hatred. It seems unlikely that neighborhood homogeneity will ever be heralded as a desired form of human settlement.

NOTE

1 Ironically, the environmental crisis of cities a century ago was remedied by separating uses (residential from industrial, for example), a strategy that later proved to generate its own set of adverse environmental effects.

4 WHY DESIGN?

In the context of city planning, design is about proposing change to urban form – streets, spaces, blocks, group of blocks, districts, or entire neighborhoods. Of special interest are the uses, locations and patterns associated with these forms.

The kind of design that is most relevant to social diversity is the kind that acknowledges the underlying social realities and possibilities of a place, a street, a block, or a neighborhood. Social dimension and complexity are relied on to generate design insight and potential. The creative role of the designer is to learn how to translate and communicate that potential. The worthiness of a project is judged by its social impact – specifically, whether it supports or undermines social diversity.

Design for diversity draws from the traditional fields of urban planning and urban design. It draws from urban planning because of the social principles planning espouses – i.e., the lessening of social inequality is central to the urban planner's code of professional conduct (see Box 1). It draws from urban design in that it is concerned with intervening in the built environment. Design for diversity merges the esthetic interest of urban design with the social objectives of urban planning. It is a way of making the urban designer's proposals more firmly rooted in social justice, and the urban planners' concern with social justice more design-based.

There are a number of reasons why planning and design of the built environment are critically important for social diversity. First, diverse neighborhoods tend to have a high number of physical transitions. Juxtapositions of difference are visible because in a diverse place there are different kinds of people doing different kinds of things. This can often be a cause of stress, particularly since the meaning and implication of various physical elements can get accentuated in diverse neighborhoods: boundaries can take on special significance, connectivity can clash with a heightened need for privacy, or visual coherence can conflict with diverse tastes and styles.

Box I American Institute of Certified Planners (AICP) Code of Ethics and Professional Conduct (emphasis added)

Our primary obligation is to serve the public interest and we, therefore, owe our allegiance to a conscientiously attained concept of the public interest that is formulated through continuous and open debate. We shall achieve high standards of professional integrity, proficiency, and knowledge. To comply with our obligation to the public, we aspire to the following principles:

a) We shall always be conscious of the rights of others.
b) We shall have special concern for the long-range consequences of present actions.
c) We shall pay special attention to the interrelatedness of decisions.
d) We shall provide timely, adequate, clear, and accurate information on planning issues to all affected persons and to governmental decision makers.
e) We shall give people the opportunity to have a meaningful impact on the development of plans and programs that may affect them. Participation should be broad enough to include those who lack formal organization or influence.
f) **We shall seek social justice by working to expand choice and opportunity for all persons, recognizing a special responsibility to plan for the needs of the disadvantaged and to promote racial and economic integration. We shall urge the alteration of policies, institutions, and decisions that oppose such needs.**
g) We shall promote excellence of design and endeavor to conserve and preserve the integrity and heritage of the natural and built environment.
h) We shall deal fairly with all participants in the planning process. Those of us who are public officials or employees shall also deal evenhandedly with all planning process participants.

Source: www.planning.org.

Second, diverse neighborhoods are often the target of policies aimed at either increasing investment or slowing down displacement. Public and private investment takes place alongside rent control, tax relief, zoning changes, or regulations on the size of new developments. Planning and design are therefore needed to help channel these policies and investments into a diversity-sustaining environment rather than one of constant conflict and tension between competing interests.

Third, design can act as a catalyst for focusing people's attention on the public realm. This is particularly important in diverse places since maintenance of social diversity requires special attention to the public realm. If neighborhood issues are framed in civic terms, residents may be motivated to think about their similarities and connections rather than their differences and conflicts. It keeps the discussion more broad, instead of focusing on particular populations (like gentrifying 'yuppies', recent immigrants, or the homeless). Design puts the public realm literally in view.

A fourth reason design is important is that inattention to design – the absence of any thought given to place quality – could undermine diversity. Social diversity is often fragile, sensitive to context. This means it can be destabilized. There are recognizable ways that the form, pattern, structure (i.e., the design) of places has thwarted the maintenance of social diversity, for example by failing to accommodate new development appropriately. As previously outlined, the tools of urban planning and design, like zoning, street standards and other kinds of regulations, have consistently played a role in undermining diversity. To reverse this requires paying more attention to how planning and design can instead be used to support diversity.

For these reasons, design strategies in diverse neighborhoods take on special significance. It's not just about putting in a new facility, having more locally-owned businesses, or developing a certain kind of housing.

Adjacent, varied housing types are a common feature in diverse neighborhoods. The photo above was taken in Irving Park, on the northwest side of Chicago.

It's about directing those efforts toward the explicit goal of supporting diversity.

DESIGN NEGLECT

It has been said that 'Every minute detail of urban design determines whether the creative geniuses in our minds are welcomed or excluded from participation in city life' (Engwicht, 2003, p. 2). Design affects all kinds of non-physical realms, things like choice, access, opportunity, interaction, movement, identity, connection, mix, security, and stability. Environmental psychologists and human geographers have documented that people are deeply affected by place, that environments can have a profound impact on human behavior and feelings (Tuan, 1981; Gallagher, 1994), that spaces can be thought of as embodied, gendered, inscribed, or contested (Low and Lawrence-Zuniga, 2003). Designed spaces are capable of conveying, reinforcing, and even legitimizing social divisions. Racial identity, for example, has a certain physical expression, tied up in things like freeways and urban renewal (Lipsitz, 2004).

Despite these known interactions, the linkage between design and social goals like diversity is often ignored. While it's right to be cautious about the relationship between social phenomena and the built environment, the translation of social diversity to principles of physical planning and design seems unnecessarily down played. Books that connect urban planning and diversity often avoid design completely. For example, Sandercock (1998, p. 5, 119) stresses that the 'normative cosmopolis', the 'utopia with a difference' is something that planners must try to evolve, but something to which she 'will not ascribe built form'. Dory Reeves' (2005) *Planning for Diversity* is focused mainly on methods of public participation, and Thomas' (1997, p. 258) approach to building 'unified diversity for social action' is predominantly a matter of making sure that planners are better educated. Texts on 'social sustainability' through diversity are likely to call for 'planning and social processes' like fiscal equality, regionalism, public transportation, and 'open and democratic local governing' (Polese and Stren, 2000). Friedmann's (2002) call for an 'open city' of diverse peoples goes somewhat further by advocating a reduction in the urban ecological footprint, chartering local citizenship, meeting basic human needs, and promoting new forms of governance, but specific physical design ideas are kept at a distance.

Given the way in which physical solutions have been cast as cure-alls throughout much of planning's history, critics are right to guard against letting planners get away with 'place' remedies at the expense of people, institutions, and political process. And yet, entire books on the benefits of neighborhood planning will include barely a mention of the critical importance of design or place (see, for example, Wright, 2001). And the usual array of recommended policies to alleviate the inequitable 'geography of opportunity' leave out the design dimension almost entirely. Policies like mixed income, fair share, and mobility housing programs are more often than not articulated in terms that do not address physical character and the design of place.

Social scientists have an interest in pointing out the connections between physical environment and social phenomena, often focusing on the strong links that can be made between social and spatial isolation (Massey and Denton, 1993). They often emphasize neighborhood as the context of social problems, from high unemployment (Granovetter, 1990), to crime (Sampson et al., 1997). But their interest is not the design of neighborhoods and cities, and when social scientists speak about the 'context' of neighborhood they are speaking about the traits of the people who live there. They may emphasize the 'political economy of place' (Logan and Molotch, 1987) or the 'social production of space' (Castells, 1983), but this excludes any specific recommendations about the design of place or space. Thus Massey and Denton (1993) repeatedly point to the fact that southern areas are often better at integration because of their 'distinctive ecological traits' (p. 73), but nowhere is physical change pursued as a target for reform. Human geographers study intently 'the importance of spatiality in the processes of social reproduction', but the spaces to be studied, the 'discourse-producing sites' like prisons, schools and hospitals are largely decontextualized (Livingstone et al., 1998, p. 145). In the social sciences, the physical environment is relied on as an explanation for social segregation, but proposed remedies steer clear of its rehabilitation.

City Planning, whose purview specifically includes the rehabilitation of the physical environment, has not spent much effort filling in this missing perspective. Witold Rybczynski recounts the history of planning's retreat from design, asserting that planning's many design mistakes – superblocks, high-rise public housing, slum clearance, government complexes – and their astounding failure caused planners to withdraw from the task of city design altogether. Planners now 'mediate, animate, negotiate, resolve conflicts, find the middle ground', which may be 'honorable', but 'it leaves the creation of an urban vision entirely to others' (Rybczynski, 2000, p. 216).

Diverse areas often have abrupt land use mixtures. This home is next to a car dealership on a main road in Blue Island, on the south side of Chicago.

Detachment from the physical context of diversity may be related more generally to the loss of localized form as a context for production and consumption – the substitution of 'flows and channels' for real places (Castells, 2003, p. 60). We consume without being affected or inhibited by the context of production, including whatever behind-the-scenes social and economic realities our consumption may require (Borgman, 1987). Perhaps under these circumstances, leveraging place to support diversity seems illogical. Or it may be that using place to support diversity seems too determinist and controlling, in danger of requiring the construction of what Harvey (1989, p. 303) calls a 'localized aesthetic image' that supports the 'capitalist hegemony over space'. Even the use of 'spatial metaphors' like 'concentration' or 'deconcentration' of poverty is seen as superficial in that it disguises the underlying political and economic processes of poverty and provides 'justification for simplistic spatial solutions to complex social, economic, and political problems' (Crump, 2002, p. 581).

By default, architects have assumed the role of linking diversity to physical structure. In SMLXL, Rem Koolhaas (1996) has devised an architecture for the multicultural city of difference, but it is a disjointed perspective employing an urbanism of 'Neitzschian frivolity' that few are likely to find appealing. Some want to glamorize 'the authenticity of disequilibrium', an urbanism pitted against the 'humanly alienating, mechanistic, equilibrium' of a city plan (Akkerman, 2003, p. 76). In what seems like a desperate attempt, there have been calls to foster diversity using 'creative disruption' as a means (Sorkin, 2006). Missing from these approaches is consideration of daily life needs, elements that are crucial to sustaining diversity in urban settings. We are left with an attempt to inspire social diversity based on disequilibrium and destruction, devoid of careful understanding of the physical elements and neighborhood structure required for everyday life.

While there is recognition that neighborhoods that support diversity must be safe and have good access to schools, employment and other services, there seems less recognition of the reality that these conditions require a concerted focus on the design of place. Design is critical in calls for promoting 'place-based initiatives' like community and economic development, worker mobility and household mobility strategies, or the reduction of service inequities. But without paying attention to neighborhood-level effects, to how these programs play out in physical terms, or how they are to be nurtured and sustained in a material context, a vitally important piece is overlooked.

In fact, many studies of deliberate attempts to create diverse neighborhoods consistently identify design as a key factor in their success. There are findings, for example, that 'the size, design, condition, location, and cost' of mixed-income housing 'are extremely important' (Ding and Knaap, 2002). Galster found that design issues were critical for sustaining mixed income housing, especially site layout, concentration, development type and scale (Galster et al., 2003, p. 175). A Massachusetts study of mixed housing type showed that tenant satisfaction did not have to do with 'subjective evaluations of neighbors', but was instead related to 'the quality of the development's design, construction, and management' (Schwartz and Tajbakhsh, 1997, p. 76, 80). In addition to addressing programmatic concerns like tenant screening, counseling, and project management, there is a corresponding need to address context, place, and design.

Connecting design to social goals like diversity may require pro-action. History has shown that neighborhood form does not always keep up with social change (Hillier and Hanson, 1984). Social transitions in the latter half of the 20th century, such as 'lifestyle and cultural diversification', women in the labor force, and smaller households were not adequately accommodated in the physical environment – and still aren't (Filion and Hammond, 2003). Now, we need an approach that can be

Commercial main streets in diverse neighborhoods are vibrant, but also struggling. Portage Park, above and West Ridge, below, are two of the diverse areas studied in this book.

responsive to a society growing more and more diverse. By mid-21st century, one-half of the population of the US is expected to be composed of today's 'minorities' (O'Hare, 1992). Planners will need to consider whether the residential structure used to house an increasingly diverse population will intensify segregation, or help to accommodate diversity.

Pursuing the objective of place diversity through the mechanisms of planning and design will require a nuanced understanding of the interconnections involved. It will require knowledge of the difference between redevelopment that contributes to loss of diversity and redevelopment that sustains diversity. To work toward stability and discourage displacement, to simultaneously support homeownership and rental housing, to successfully integrate a range of housing types and densities, levels of affordability, a mix of uses, and neighborhood facilities and social services – all of this together requires holistic attention that includes the physical form and design of neighborhoods.

PART TWO

THE CONTEXT

In this part, I lay the groundwork for the design strategies contained in Part III. Using the City of Chicago and its surrounding suburban areas as a case study, I investigate whether social mixing is related to particular patterns and structures found within the urban environment. I select a measurement approach, spatial unit and time period, and I look at geographic patterns of diversity, the associated neighborhood forms, and some explanatory factors that can be used to predict these patterns and forms.

I then present a summary of 85 interviews of residents who live or work in six of Chicago's most socially diverse places. The point of these interviews is to try to understand how residents feel about social diversity, what they consider to be positive and negative aspects of diverse neighborhoods, what challenges they see, and how design plays a role in their neighborhood's improvement and stabilization.

5 PATTERNS

Social diversity is geographically patterned. It can be mapped in various ways, with different conclusions drawn, depending on how diversity is defined and measured. This chapter looks at the patterns of diversity in the City of Chicago and the surrounding suburbs and towns

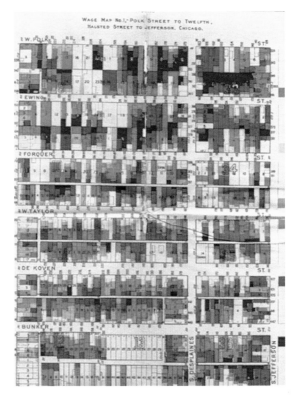

Nationalities Map No. 1 – Polk Street to Twelfth, Halsted Street to Jefferson, Chicago, IL. Source: Hull-House Maps and Papers, by residents of Hull-House.

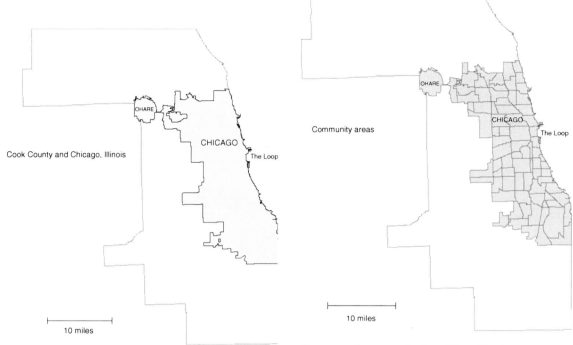

Study area: Cook County, which includes the City of Chicago.

Community areas in the City of Chicago, most of which were delineated in the 1920s by Chicago School sociologists.

of Cook county (see above). The County consists of 138 incorporated towns, 866 census tracts, and had a 2000 census population of 5.3 million (making it the second largest county in the US). Within the City of Chicago, there are 77 'community areas' and 172 officially designated neighborhoods.

MEASURING DIVERSITY

When people talk about 'diversity', they sometimes mean 'ethnic' or 'minority'. As if to say that if one is studying a predominantly African-American or Hispanic neighborhood, one is therefore studying 'diversity'. In fact, and unfortunately, many African-American neighborhoods in Chicago are alarmingly not diverse – they are highly segregated.

Here, diversity is defined as the mix of different groups within a neighborhood. Groups are defined along four dimensions: age, income, family type and race/ethnicity. These correspond to the normative ideals attached to social mixing discussed in Chapter 3. The goals of vitality, economic growth, tolerance, sustainability and social justice are accomplished through the mixing of races, ethnicities, ages, family types and income levels. There may be other important dimensions to consider (such as education level), but these particular four are fairly comprehensive.

Evaluating neighborhood-level social mix is sometimes approached by way of segregation indices. Focusing on segregation rather than integration is an indirect measure, and some argue that this is causing researchers to harbor biases against integrated areas (Smith, 1998). Social segregation is measured by socioeconomic status, stage in the life cycle, or race and ethnicity, although most information we have about segregation is devoted to racial segregation.

Dimensions of spatial segregation can be looked at in terms of isolation vs. exposure, and evenness vs. clustering. Most well-known are the indices of dissimilarity and isolation that have been used to evaluate residential segregation levels. Massey and Denton (1986) included five dimensions of racial residential segregation – evenness, exposure, concentration, centralization, and clustering – each corresponding to a different measure. However, the dissimilarity and isolation indices have been critiqued for failing to account for spatial location, for ignoring the critical importance of surrounding land usage, and for having an anti-urban bias (Quinn and Pawasarat, 2003).

Measures of integration, as opposed to segregation, present a different set of measurement issues. An important distinction must be made between measures that are comparative, where the proportions in a sub-unit (a neighborhood) are compared to those in a larger unit (the entire metropolitan area), versus measures that are absolute, where a predetermined statistic defines integration. Most integration studies evaluate the proportion of one group relative to another (Clark, 1993; Lee and Wood, 1990). Galster (1998) defined a neighborhood as 'mixed' if no single group made up more than 75% of the population. Quinn and Pawasarat (2003) labeled a neighborhood 'Black-White integrated' if the population was at least 20% Black and 20% White. Ellen (1998) defined 'racially integrated' as neighborhoods with a Black population between 10 and 50%, which she justified as a compromise between the fact that Blacks make up a relatively small percent of the population (roughly 13%), and the idea that space should be shared equally.

A key issue is how to define the stability of integration. Saul Alinsky famously remarked that integration is the time between the first Black moving in and the last White moving out (Sanders, 1970). Clearly, a neighborhood in the process of transitioning from one group to another cannot always claim to be reaping the benefits of diversity (this is the case with land use diversity as well, and Jacobs wrote about it in the context of one use becoming overly dominant in one location). Neighborhoods may either be in the process of losing particular groups (White flight), or, one group may be displacing another (gentrification). Strategies for measuring the stability of integration or diversity include comparing how the proportion of one group relative to another changes over time (Galster, 1990), and then determining a threshold or 'tipping point' that would indicate flight or displacement. In a comparative approach, changing proportions would be evaluated in function of overall citywide population shifts. In a market approach, stability would be a matter of assessing population proportions relative to housing affordability (Smith, 1998). Galster's (1998) 'stock/flow model' defines stable integration on the basis of maintaining some diversity in the range of household flows in and out of the neighborhood over a ten year period.

Those focusing more specifically on diversity than integration and segregation may use multigroup methods rather than successive two-group combinations. For example, Maly (2000) used a 'neighborhood diversity index' to compare four population groups. Nyden et al. (1997) also used a comparative approach (rather than an absolute definition of diversity), and defined a diverse census tract as one that approximated the racial and ethnic composition of the city as a whole. Keating (1994) considered a neighborhood integrated if its population was 5–39% non-White. But Ellen (1998) pointed out that the use of the comparative measure has a serious downside since it means that overwhelmingly homogeneous neighborhoods could be labeled integrated if their larger communities were similarly lacking in diversity.

Various authors have attempted to define different kinds of residential mix, and this has resulted in a number of proposed typologies. Khadduri and Martin (1997), for example, used three categories of mix to define 'mixed-income housing' involving HUD public housing projects. Smith (2002) proposed five categories based on the kind of income mix involved (e.g., 'low-income inclusion' or 'broad range of incomes'). Immergluck and Smith (2003) classified neighborhoods in Chicago according to the internal mix of high and low-income residents, ranging from 'highly restrictive' to 'highly diverse'.

Diversity measures have also been used extensively in the biological and environmental fields. A paper by Maignan *et al.* (2003a), 'Bio-Ecological Diversity vs. Socio-Economic Diversity: A Comparison of Existing Measures' identified seven separate diversity measures in the bio-ecological fields that could be applied to socioeconomic data. Two of the most common measures in the biological field are the Hurlbert (1971) and the Simpson (1949) indices, where perfect diversity is defined as a uniform distribution among all categories. A unique study by Byrne and Flaherty (2004) used biological diversity indices to look at whether the housing market (types of dwellings and types of occupants) was becoming more or less diverse. Land use diversity can be evaluated similarly, using biological measures to evaluate how many land use categories exist in a given spatial area.

While there are many measurement approaches to choose from, I selected two to analyze the diversity of Chicago: the Simpson Diversity Index and the Neighborhood Diversity Index. The Simpson Index has been around since the 1940s. It can be used to evaluate how many categories (e.g., income levels, races, ethnicities, housing types) exist in a given area. The index is commonly used to measure biological diversity. Its formal expression is:

$$A = \frac{N(N-1)}{\sum i \, n_i(n_i - 1)}$$

where A is the diversity index; N is the total number of individuals (or housing units or households) for all categories; and n_i is the number of individuals (or other characteristic) in the ith category.

I looked at four variables to characterize resident diversity: income, race/ethnicity, age, and family type, listed in Table 5.1. Also listed are a few 'residential variables' to characterize the diversity of other factors, for example, the diversity of housing unit types and sizes. Table 5.1 also lists the parameters used to define groups within each variable.

The data comes from Census Summary Tape File 3, spatially adjusted to allow comparisons between geographic areas over time.[1] The geographic level used is the census block group, which provides an adequate level of spatial variability without being too aggregated. Although the census tract is often used in studies of racial and class segregation, researchers have argued that smaller units may be more appropriate.[2]

Table 5.1 Description of variable categories, used to compute diversity indices

Diversity Variable	Categories
Race/Ethnicity	White alone/Black alone/ Asian alone or Pacific Islander alone/Hispanic/Other
Age	5 and under/6 to 18 years/19 to 34 years/35 to 64 years/65 and over
Family Income	Under $20,000/$20,000 to $39,999/$40,000 to $74,999/$75,000 and over
Family Type	Married, with children under 18/ Married, no children under 18/ Single, with children under 18/Single, no children under 18/Non-family household
Housing Unit Type	1 unit detached/1 unit attached/2 units/3 or 4 units/5-9 units/10-19 units/20-49 units/50+ units
Housing Tenure	Owner occupied/Renter occupied
Year Built	Built 1939 or earlier/Built 1940-1959/Built 1960-1979/Built 1980 or later
Unit Size	No bedroom/1 bedroom/2 bedrooms/ 3 bedrooms/4 bedrooms/5+ bedrooms
Housing Value	Less than $100,000/$100,000 to $174,999/$175,000 to $299,999/$300,000 and over
Monthly Rent	Under $500/$500 to $799/$800 to $1,249/$1,250 and over

Note: All variables are from the 2000 Census, by block group.

SPATIAL PATTERNS

The figures in the next page show the distributions of the most diverse block groups for four variables – income, race/ethnicity, age, and family type. The following figure shows their combined distribution. The maps show that the most diverse areas are generally located in the inner-ring, pre-World War II suburban areas. Despite the increasing tendency of newer suburban areas to intensify and become more diverse, the more suburban (outlying) areas in Cook County do not appear to be as diverse as the inner-ring locations.

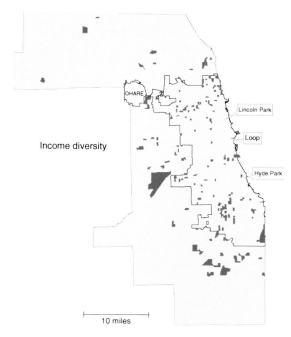

The most income diverse areas in Cook County.

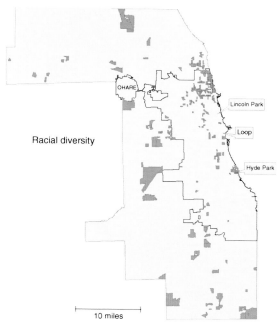

The most racially/ethnically diverse areas in Cook County.

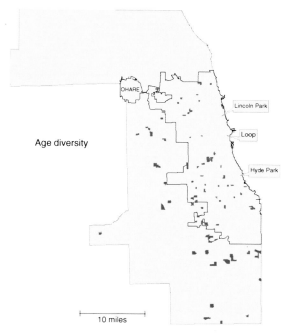

The most age diverse areas in Cook County.

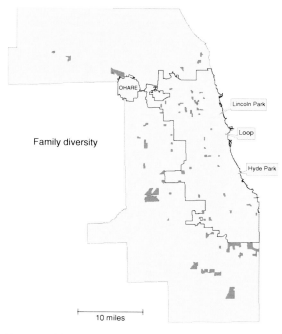

The most family type diverse areas in Cook County.

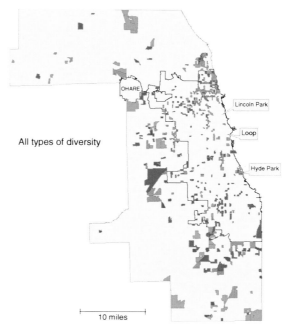

All types of diversity

Most diverse on four measures: income, race/ethnicity, age and family type.

The next two figures compare areas with high-income diversity to two land use characteristics – industrial land and housing age. The first figure shows the block groups in the highest quartile of income diversity together with land areas classified as industrial. There appears to be a definite spatial connection – high-income diverse areas are adjacent to industrial land. There is also a relationship between high-income diversity and housing age, although it is less pronounced. The second figure shows those block groups with a median housing age of 1941 or earlier (the earliest age quartile), and those block groups in the highest quartile of income diversity, constituting 339 block groups.

The picture that emerges, then, is that block groups in the top quartile of income diversity represent both the 'classic' Chicago neighborhood of the 1870s–1920s, the inner-ring areas of Chicago, as well as some industrial suburbs in the south and southwest. Many of these areas are not only income diverse but traditionally multiethnic. Their urban form is generally comprised of low and mid-rise structures adjacent to commercial street corridors. The commercial structures line

The spatial proximity between industrial land and areas with high-income diversity.

thoroughfares connected to downtown, with ground floor retail, housing or offices above, and single-family bungalows and duplexes on relatively small lots on the adjacent residential streets. There are also mixes of unit types (bungalows next to small apartment buildings). Although many of the commercial areas are thriving, there are examples of commercial areas with vacant lots and poorly defined open space. The following four figures show representative images taken from these diverse areas.

High-income diversity and pre-1942

High-income diversity

pre-1942

0   3   6   12 Miles

N W E S

The spatial proximity between older housing stock and areas with high-income diversity.

INCOME DIVERSITY AND NEIGHBORHOOD TYPE

What can be said more definitively about the neighborhoods in which diverse block groups are located? To answer this, I selected a definition of neighborhood type, and then determined the neighborhood types diverse block groups fell into. Neighborhood types were defined using a typology devised by staff at the University of Chicago Map Library.[3] Their neighborhood types were created by first factoring out four dimensions from 34 census variables, and then performing a

New condominium development is common in diverse neighborhoods (Bridgeport).

Mixtures of housing types are common in diverse neighborhoods (West Ridge).

Diverse neighborhoods are often located close to major transportation corridors and other strong edges (Bridgeport).

Diverse neighborhoods are still holding on to small independent retailers (West Ridge).

cluster analysis of these four dimensions. This yielded 10 neighborhood types, which are listed at the bottom of Table 5.2. The types vary from being, for example, very urban and impoverished, to very suburban and wealthy. Neighborhoods are equivalent to census tracts.

Table 5.2 High-income diversity block groups and neighborhood type

Neighborhood Type (see Table 5.4)	Count (number of block groups in top income diversity quartile)	Percent
1	159	15.00
2	236	22.26
3	192	18.11
4	46	4.34
5	181	17.08
6	149	14.06
7	4	0.38
8	83	7.83
10	10	0.94

Just focusing on income diversity, the table shows that most income diverse block groups fall into neighborhood types '2', '3' or '5': The greatest number of diverse block groups are located in the neighborhood type defined as African-American, somewhat impoverished, and on the edge of more impoverished type-1 neighborhoods. The second most common are neighborhoods described as blue-collar and ethnic, located in the inner suburban or outer city areas. The third most common neighborhood type is described as 'inner-city Hispanic' (type '5'). No block groups were located in 'suburban, well-off' neighborhoods, and only 0.38% were located in 'urban, very well-off' neighborhoods.

Is it only income diversity that can be associated with these types of neighborhoods? To answer this, I wanted to explore how the results changed if a broadened, more multi-dimensional definition of diversity was used. I used four alternative definitions (see Table 5.3). First, block groups that ranked high (in the top quartile) on all four resident diversity indices (income, race/ethnicity, age, family type) were classified as areas with high 'social diversity'. There were 49 block groups that met this condition. Second, block groups that scored high on residential

Table 5.3   Neighborhood characteristics for high diversity block groups

	Neighborhood Types									
	1	**2**	**3**	**4**	**5**	**6**	**7**	**8**	**9**	**10**
Social Diversity: Block groups in the top quartile on all four social diversity variables: income, race, age, family type. (49 block groups)		9	23		11	3		3		
Residential Diversity: Block groups in the top quartile for diversity of units, tenure, year built, and no. bedrooms, and top two quartiles (above the median) for housing value and rent diversity. (20 block groups)	5	1	7	1	3	2			1	
Social and Residential Diversity: Block groups in the top quartile on all four social diversity variables: income, race, age, household type; plus, top quartile for unit diversity and top two quartiles for housing value and rent diversity. (10 block groups)		1	5		4					
Stable Diversity: Block groups in the top quartile in 1990 and 2000 for income and age diversity, and unit and tenure diversity. (7 block groups)			4							

variables (unit type, year built, etc.) were classified as having high 'residential diversity'. This condition applied to 20 block groups. A third category, 'social and residential diversity' was used to define an even smaller subset of 10 block groups, consisting of a combination of social and residential diversity variables. Finally, block groups that exhibited 'stable diversity', which characterized only seven block groups, was defined as high diversity for income, age, unit and tenure for both 1990 and 2000. Note that no block groups were in the top quartile on all measures of diversity for both years. However, there was some overlap – some block groups were included in more than one category used to define block group diversity.

Even with these broadened definitions, Table 5.3 shows that most diverse block groups still fall into neighborhood types '2', '3' or '5'. The greatest number of block groups was located in the neighborhood type defined as blue-collar and ethnic, located in the inner suburban or outer city areas (Table 5.4). The second most common neighborhood

Table 5.4 Explanation of neighborhood types – Tables 5.2 and 5.3

1	Very urban, impoverished, English-speaking, with many female-headed families and numerous children. The core impoverished African-American neighborhoods of the South and West Sides.
2	Somewhat impoverished, mostly English-speaking, with a fair number of female-headed families with many children. Mostly African-American neighborhoods on the edge of type-1 neighborhoods.
3	Somewhat urban and somewhat linguistically-isolated. Mostly blue-collar, often somewhat 'ethnic' neighborhoods in the outer city and inner suburbs.
4	Very well-off neighborhoods with many non-family households. Most of the North Side Lakefront, plus the area around the Loop, with outliers in Hyde Park, Evanston, Oak Park, and a few suburban tracts with apartment building clusters.
5	Urban, impoverished, and very linguistically-isolated/Hispanic (more than 2.5 standard deviations above the mean on the latter). Inner-city Hispanic neighborhoods.
6	Very urban and very linguistically-isolated/Hispanic, with non-family households. The complicated, often only partly Hispanic, neighborhoods on the inner Northwest and Far North Sides.
7	Urban, very well-off, with a great many non-family households (nearly 4 standard deviations above the mean on the latter). Neighborhoods with numerous young, unmarried adults and hardly any children.
8	Not especially wealthy. The outermost suburbs, the inner southwest suburbs, and much of Northwest Indiana.
9	Suburban, well-off. More prosperous suburbia. Concentrated especially in the western and northwestern suburbs.
10	Very suburban, very wealthy, mostly English-speaking. Highly prosperous suburbia; more than 2.5 standard deviations from the mean on wealth.
11	Very urban, impoverished, English-speaking, with many female-headed families and numerous children. The core impoverished African-American neighborhoods of the South and West Sides.
12	Somewhat impoverished, mostly English-speaking, with a fair number of female-headed families with many children. Mostly African-American neighborhoods on the edge of type-1 neighborhoods.
13	Somewhat urban and somewhat linguistically-isolated. Mostly blue-collar, often somewhat 'ethnic' neighborhoods in the outer city and inner suburbs.
14	Very well-off neighborhoods with many non-family households. Most of the North Side Lakefront, plus the area around the Loop, with outliers in Hyde Park, Evanston, Oak Park, and a few suburban tracts with apartment building clusters.
15	Urban, impoverished, and very linguistically-isolated/Hispanic (more than 2.5 standard deviations above the mean on the latter). Inner-city Hispanic neighborhoods.
16	Very urban and very linguistically-isolated/Hispanic, with non-family households. The complicated, often only partly Hispanic, neighborhoods on the inner Northwest and Far North Sides.
17	Urban, very well-off, with a great many non-family households (nearly 4 standard deviations above the mean on the latter). Neighborhoods with numerous young, unmarried adults and hardly any children.

(Continued)

Table 5.4   (Continued)

18	Not especially wealthy. The outermost suburbs, the inner southwest suburbs, and much of Northwest Indiana.
19	Suburban, well-off. More prosperous suburbia. Concentrated especially in the western and northwestern suburbs.
20	Very suburban, very wealthy, mostly English-speaking. Highly prosperous suburbia; more than 2.5 standard deviations from the mean on wealth.

Source: University of Chicago Library, http://www.lib.uchicago.edu/e/su/maps/chi2000.html

type is described as 'inner-city Hispanic' (type '5'), followed by African-American neighborhoods on the edge of the most impoverished African-American areas (type '2'). Only one block group was located in a 'suburban, well-off' area, and this block group contained a cluster of some low-rise apartments together with single-family dwellings units – thus scoring high in terms of residential diversity. Only one block group was located in the 'very well-off' neighborhood type, the North Side Lakefront area, and this too was a matter of residential as opposed to resident diversity.

DENSITY, UNIT TYPE AND DIVERSITY

Finally, a few thoughts about the relationship between density, unit type and diversity. In a previous study using regression analysis, I found that diversity and density were positively correlated – i.e., for the vast majority of block groups, higher density predicted higher diversity (Talen, 2006). The peak density point was 57,239 persons per square mile.[4] While block group density can top 100,000 per square mile in Chicago, the peak density point was in a fairly high range, since only 30 block groups in Cook County have a density higher than 57,000. Table 5.5 lists the unit characteristics associated with eight block groups in this density range, along with the neighborhood each is located in, and the income category distributions. The table indicates that high diversity is sometimes associated with large apartment buildings, (in buildings of 20 units or more), but sometimes it is characterized by groups of buildings with fewer units (duplexes and small apartment buildings). The neighborhoods these block groups are located in have a reputation for high racial/ethnic diversity. Edgewater, for example, is a neighborhood on the far north side that was only 48% non-Hispanic White in 2000.

Table 5.5 Unit characteristics of eight high density, income-diverse block groups

Block Group	Neighborhood	1-unit detached	1-unit attached	2 units	3-4 units	5-9 units	10-19 units	20-49 units	50+ units	Income Cat. 1	Income Cat. 2	Income Cat. 3	Income Cat. 4
170310301002	Edgewater	42	14	11	0	29	46	450	826	135	105	80	62
170310306001	Edgewater	34	24	15	38	127	88	500	2136	294	218	292	226
170310315003	Uptown	0	0	6	39	160	248	167	160	167	67	46	27
170310715002	Edgewater	33	136	28	85	148	83	132	1608	0	33	33	388
170310812002	Near North Side	30	19	0	0	33	32	98	1385	14	41	39	397
170311401002	Albany Park	17	7	147	184	160	90	63	4	133	175	135	65
170313009001	South Lawndale	40	17	264	362	173	31	0	7	239	274	165	36
170313109002	Lower West Side	61	17	196	193	198	0	0	0	109	189	99	47

Selected block groups are closest to the peak density of 57,239 persons per square mile (56,000–58,400), Census 2000. Income categories refer to number of families in each category listed under Family Income in Table 5.1.

It's important to note that areas with the highest density are not the most diverse, a point not surprising to anyone who has witnessed the rapid growth of luxury high-rise condominium development in downtown Chicago. Still, the results do *not* support the conclusion that lowerdensity suburbs are strong on social diversity. On the contrary, the results suggest that the areas that are the most diverse are in the 'inner-ring', corroborating other studies that have shown that middle-income, first-tier suburbs have the widest range of income groups (Oliver, 2001).

The regression analysis cited above also concluded that a mix of housing types does not always predict income diversity. This means that unit diversity is not likely to be sufficient as a means of promoting diversity. The study did find an association between diversity and housing value mix, owner/renter mix, and a mix of housing ages combined. While we can't rely on unit mix to create socially diverse places, unit diversity in conjunction with other types of diversity – housing value, tenure and age – may have a significant effect.

One explanation for the lack of clearer association between unit diversity and social diversity may have to do with the desirability of older neighborhoods. Affluent areas in the north of Chicago are characterized by a historical mix of units over ground-level retail, townhouses, and single-family homes, now occupied by people willing to bid high for the price of living in an early 20th century traditional neighborhood. Such areas may have been occupied by a broader mix of income levels as their traditional form and pattern warranted, but this diversity has been lost over time because of a scarcity of similar neighborhood qualities. This transformation does not occur in all older neighborhoods, however. In some poorer neighborhoods, the mix of unit types may not be accompanied by an adequate level of services and public facilities necessary to attract (or retain) higher-income residents.

NARROWING THINGS DOWN: SIX HIGHLY DIVERSE PLACES

To select places for in-depth study, I needed a method for narrowing down the diversity of Chicago's neighborhoods into a smaller set of areas. My strategy was to use two measures, the Simpson Diversity Index and the Neighborhood Diversity Index (Maly and Leachman,

1998). The latter index compares the population distribution of an area to the overall city average. Its formal expression is:

$$ND = \frac{1}{2}\left(|Ca - Ta| + |Cb - Tb| + |Cc - Tc| + |Cd - Td|\right)$$

where C is the group percentage (categories a, b, c, d, for example, using racial or income categories) for the whole city, and T is the group percentage for the area, such as a census tract. An area reflecting the composition of the city will have a low ND, while an area consisting predominantly of one group will have a large ND.

Combining these two measures provides a degree of robustness — places that are highly diverse on both measures are more likely to be diverse no matter how diversity is measured. I mapped those census block groups that were the most diverse — scoring in the top quartile on both measures — for family type, income, age, and race/ethnicity.

To make a final selection, I first selected block groups that were in the top quartile on income diversity for the Simpson index and the ND index in both 1990 and 2000. I wanted to focus first on income diversity because I think it's the most important and relevant measure of diversity when it comes to design — i.e., design strategies like unit mix can have an effect on income diversity in a way that is more direct than the other diversity types. I excluded any block groups that had 0 families in either year, leaving a total of 4,242 block groups. Out of those, 165 block groups were in the top quartile ranking of diversity on both indexes in both years.

I then selected block groups that were diverse on income plus one other variable (age, family, or race/ethnicity); seven were income diverse and age diverse; seven were income diverse and family diverse; and 21 were income diverse and race/ethnic diverse. I then looked for spatial clusters of these block groups. I found six areas that had clusters of three or more block groups within one community area or one municipality. These are shown in the next figure.

These six areas — the incorporated towns of Berwyn and Blue Island, and the community areas of Bridgeport, Irving Park, Portage Park and West Ridge located in the City of Chicago, were the selected locations for the resident interviews. The results of these interviews are the subject of Chapter 6.

The four community areas in Chicago — Bridgeport, Irving Park, Portage Park, and West Ridge — all had good datasets for investigating physical

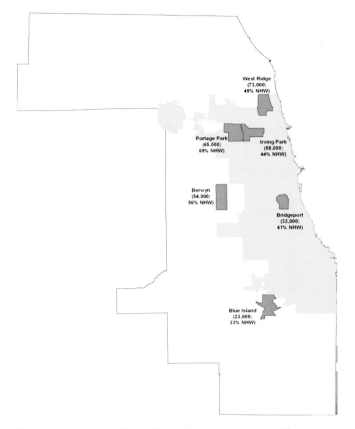

These six areas in Cook County were selected for
further study because they are highly diverse on
multiple dimensions.

form (parcels, building footprints and heights, land use, etc.). These
areas were therefore selected for the design studies – the subjects of
Chapters 7, 8 and 9.

NOTES

1 'Spatially adjusted' means that the boundaries between census geographic
   units, which change over time, have been spatially rectified so that time
   periods can be meaningfully compared. This is done by spatial interpola-
   tion, whereby data is apportioned to new spatial units. The disadvantage
   of using Summary Tape File 3 is that, since the data is based on a sample
   rather than a full count, it is subject to sampling error. For this study, I felt
   that the problem of sampling error was outweighed by the need for a small

enough geographic unit for gauging diversity (block group), and the need to look at change over time.

2 Quinn and Pawasarat (2003), for example, noted that racial integration that relies on aggregate measures like census tracts may be misleading, advocating instead for block-level measurement.

3 The data was obtained from The Joseph Regenstein Library at the University of Chicago. Maps from their website as well as a detailed explanation of their methodology can be viewed at http://www.lib.uchicago.edu/e/su/maps/chi2000.html.

4 At higher densities, diversity declined, i.e., the relationship between density and diversity became negative.

6 THE INTERVIEWS

From February to June, 2006, tape-recorded interviews were conducted of 85 residents in the six neighborhoods identified in Chapter 5 as being 'highly diverse' on multiple dimensions (see Table 6.1 for census statistics of the six neighborhoods). Six students were hired to help conduct the interviews. Interviewers used a semi-structured format, starting with a list of questions (Box 2) that respondents were then encouraged to expand on. The following four issues were of primary interest:

1 Familiarity with, and opinions about, social diversity.
2 Social bonds and connections in the neighborhood.
3 Place-identity – whether respondents identified with *place*, and what features seemed to promote that identity (e.g., a special restaurant, the place they grew up in, a street, a favorite park).
4 Strengths and weaknesses of the neighborhood/community.

We started with respondents we could easily connect to: local business owners, community center representatives, chamber of commerce directors, school principals, and aldermen. Following each interview, we asked for referrals. An effort was made to identify people from all walks of life, from the public, non-profit and private sectors, both officials and residents with no official role in the community, long-timers and recent arrivals, and people from varying age groups and cultural and ethnic affiliations. Table 6.2 gives a few characteristics of the respondents. Because we often started with contacts associated with particular organizations rather than 'man on the street' interviews, the respondents tended to be people who were committed, knowledgeable and involved in the community.

This chapter starts with a brief summary of the key issues and defining characteristics associated with each of the six neighborhoods, followed by a summary of the interview responses. The summary condenses the material into a series of specific, recurring themes. One note of clarification: the terms 'community', 'neighborhood' and 'town' tended

Table 6.1   Selected Census Statistics for the Six Survey Areas*

	Population	Population per square mile	Percent White	Percent Hispanic 1990	Percent Hispanic 2000	Percent Black or African-American	Percent Asian	Percent 15 years or younger	Percent 65 years or older	Median house-hold income ($)
Blue Island	23,463	5,822	53.7	24.5	37.9	24.1	0.4	25.4	9.8	36,520
Berwyn	54,016	13,876	73.4	7.8	38.0	1.3	2.6	22.4	13.5	43,833
Bridgeport	33,694	16,044	41.0	26.1	30.2	1.05	26.1	20.3	10.7	35,535
Irving Park	58,643	18,156	44.2	24.4	43.3	1.91	7.4	20.0	9.0	42,037
Portage Park	65,340	16,417	69.5	7.8	23.0	0.51	3.0	22.4	18.4	45,117
West Ridge	73,199	20,736	49.7	8.3	15.5	6.78	22.3	19.9	13.6	41,144

*2000 data unless otherwise noted

Box 2   Interview questions

1 Please give your name and your affiliation (occupation or membership, if applicable).

2 How long have you lived in this area and how familiar are you with it? Are there particular locations that you are most familiar with? Are there special organizations, clubs, religious or community groups you are involved with?

3 How familiar are you with the social diversity of this community?

4 Do you think there is a healthy sense of community or civic life here? Do different types of people interact with each other in positive ways? Are all groups involved, or is there some disenfranchisement?

5 Are there ways in which an outsider would be able to tell, just by walking down the street, that this is a socially diverse place? (See if respondent can point out places on a map.)

6 Have there been any tensions that have resulted from the diversity of people here? Does this area embrace all types of people, including African-Americans, Latinos, Asian-Americans, immigrants? Rich and poor? Young and old? People with different types of living arrangements (family and non-family)? Have tensions been expressed in public settings, like in schools?

7 Are there ways that people here have tried to establish barriers between groups, both physical and non-physical? (For example, putting up fences and gates?)

8 Conversely, do you think the community sustains diversity in specific ways?

9 What are some of the important issues here in terms of planning/design? What kinds of changes have been occurring? (For example, are there problems with gentrification/displacement, or disinvestment?)

10 Can you think of ways that the physical structure of this community either sustains or detracts from social diversity? For example, do you see any of the following as problematic? (refer to map):
- Mixtures of different types of housing (apartments near single-family residences, for example)
- Mixtures of uses (too much or not enough, or not well designed)
- Public facilities like parks, schools, community centers (is public space sufficient and well placed – accessible to everyone?)

Box 2   Continued

- Uses that are degrading to the community – like highways, vacant land, or noxious facilities?
- Streets and intersections? Can people get around easily?

11 Can you think of specific types of policies, developments, or projects (private or public) that would specifically enhance or preserve social diversity?

12 Do you have any additional thoughts on the issue of social diversity, in Chicago or in the US more generally?

13 Finally, can you suggest other people who would be good to talk to about this topic?

Table 6.2   Respondents

Total = 85

Location	West Ridge	15
	Irving Park	15
	Portage Park	15
	Bridgeport	10
	Blue Island	12
	Berwyn	18
Race/Ethnicity	African-American	6
	White	53
	Latino	10
	Asian	8
	Unknown	8
Gender	Male	45
	Female	40
Occupation	Architect/planner/developer	10
	Business owner	17
	Community-based institution	25
	Other/unknown	33

to be used interchangeably in discussions with respondents. Technically, West Ridge, Bridgeport, Irving Park and Portage Park are 'community areas', and Berwyn and Blue Island are separate, incorporated municipalities. Historically, all were connected to rail lines, but their origins were different: Blue Island started as an agricultural center, Bridgeport as an industrial town, and Berwyn, Portage Park and Irving Park were commuter rail suburbs (Keating, 2005). West Ridge was the western-most part of Rogers Park, a commuter rail suburb, but its origins were also tied to agriculture. Within each area, there are distinct neighbor-hoods, ranging from few to many. We did not attempt to 'correct' peo-ple in their use of the terms 'community', 'neighborhood' or 'town', so the summations below tend to reflect a correspondingly descriptive rather than technical use of terminology.

A FEW PERTINENT FACTS

Most of these communities have entire books written about them, especially their histories. The books written on Chicago and its people and neighborhoods could fill a library. Sketched out below are just a few pieces of information of particular relevance to the study of each area's social diversity.[1]

West Ridge, or West Rogers Park, 8 miles north of downtown Chicago ('the Loop'), was a farming community up until the 1920s, when developers started to buy up farm land and build bungalows and apartment buildings. After World War II, there was an enormous building boom in West Ridge. Businesses in West Ridge in the 1940s, 1950s and 1960s were almost entirely Jewish owned and were clustered along the main commercial corridor, Devon Avenue. Sometime in the 1970s, 'for whatever reason' as one business owner explained, aging Jewish business owners started to move out (there were six Kosher butchers on Devon Ave at one time; now there is one), and Indian and Pakistani businesses moved in. The businesses along Devon Avenue are now predominantly Indian and Pakistani, and people come from all over the US to visit this distinct commercial area. The businesses are mutually reinforcing. People shop, find jobs, eat Indian food, plan a trip, buy wedding clothes – engage in anything related to Indian/Pakistani culture.

In recent years, West Ridge has become more diverse in religious, cultural and ethnic terms, and one respondent claimed that it now contains over 60 different religious organizations. The population of West Ridge remains strongly Jewish, even if the business establish-ments have substantially decreased. Conversely, many Indian and Pakistani business owners do not live in West Ridge, and commute in from the suburbs to run the shops, as they did initially in the 1970s.

In West Ridge, affordable apartments over stores on the main commercial corridor, Devon Avenue, are home to recent immigrants.

There is a large Orthodox Jewish community in West Ridge. Some Jewish businesses remain, but most of the commercial area is now Indian and Pakistani.

West Ridge apartment buildings have been converting to condominiums, as in many other parts of Chicago.

Some respondents remarked that most residents were detached from the main commercial activity, and in fact avoided it. As one put it: 'If you live here, you avoid Devon ... the street is crazy'. The disconnect between the commercial hub and the residents of West Ridge is palpable. Cultural diversity along Devon Avenue is driving the commerce and tourist trade of the area, but it is not servicing local residents.

Irving Park   located 7 miles northwest of the Loop, is divided between Old Irving Park, a wealthier, mostly single-family area with large historic homes, and Irving Park proper, which is a highly diverse, although strongly working-class section of Chicago. Irving Park is populated by many people of Eastern European descent. Old Irving Park has a strong neighborhood association that has been successfully blocking development in recent years, especially new multi-family developments. Many sections of Old Irving Park were downzoned to R-2 in the 1990s.

Irving Park began as a separate, affluent suburb in the 1880s but, according to the president of the Irving Park Historical Society, immigrants began building smaller homes in-between the wealthy already in the early 20th century. There may have been 'some envy about the

Streets lined with 'two-flats' and 'three-flats' are common in Irving Park.

mansions', but there were few overt tensions. Over the past two decades, Irving Park has grown more Hispanic (from 9 to 43% Hispanic in 20 years) and has become less of an Eastern European enclave. The population also became much younger, and housing values have soared.

Portage Park lies immediately to the west of Irving Park. Unlike Irving Park's division into two distinct areas, Portage Park is thought of as one neighborhood. It is not considered to be as strongly ethnic as other parts of Chicago (there are no officially organized ethnic associations), although there is a significant Polish and, increasingly, Hispanic population. Economically, many view it as a blue-collar neighborhood that has recently been settled by professionals moving out of more expensive locations like Lincoln Park.

In addition to a large centrally located park for which the neighborhood is named, Portage Park is especially proud of a once-thriving commercial area known as 'Six Corners'. This intersection of three major streets was once the most important retail location outside of the Loop. It still retains the original Sears store, which is a source of great community pride. However, the shopping area has been deteriorating over the past few decades, and many residents expressed

The intersection known as 'Six Corners' in Portage Park. This area used to be a major shopping destination for Chicago, as well as a source of great pride for local residents.

Carl Schurz High School in Portage Park is a landmark building, begun in 1908. It's now the 3rd largest high school in Illinois, and students are bussed in from areas outside the neighborhood.

Bungalows like these in Portage Park are common in all of the diverse neighborhoods studied here.

sadness at the loss of its prominence as the community hub of Portage Park. Some blamed the deterioration of Six Corners on increased car dependency; others blamed it on mismanagement and the corruption of local leadership.

Bridgeport, just south of the Loop, was the first neighborhood in Chicago, established to house laborers building the canals. It is well-known as the home of the Irish, the Daley family, and the Chicago White Sox. It also has a large planned manufacturing district which was the first of its kind in the US, set up to retain an industrial employment base. There is continuous pressure from developers to turn the beautiful old factory buildings into loft space and condos, but the city is intent on retaining the buildings for industry.

More than any of the other communities, Bridgeport has seen an explosion of condominium and high-end single-family development in the last few years. A large new complex of single-family residences called 'Bridgeport Village' has been constructed just north of the planned manufacturing district. Long-time residents of Bridgeport seemed amused by the development because of its proximity to what locals call 'bubbly creek', an odorous section of the Chicago channel where they used to dump animal carcasses. One resident said:

Older housing stock in Bridgeport is a living reminder of its blue-collar roots.

New condos are a common sight in Bridgeport. Residents complain that the new buildings do not integrate well with the surrounding neighborhood.

'Bridgeporters are laughing at them ... when you see people spending a million dollars to live next to a smelly creek, you think they're crazy'.

Blue Island is a small town (about 23,000 population) that lies just outside the Chicago city limits, on its southern border. There is a rail stop on the Metra – the regional railway serving six counties – that stops right along Blue Island's historic main street. Both the main street and the rail stop are the target of revitalization efforts, and plans are underway to transform the area into a mixed-use, transit-oriented development. Appropriate to that goal, almost 50% of Blue Island's housing stock is in the form of multiple-family dwellings.

The population of Blue Island is almost evenly distributed between African-American, Hispanic, and White residents. One resident pointed out, however, that this is not the distribution in the schools, which have much higher numbers of Hispanic children. This reflects recent in-migration. During the 1990s, thousands of Whites moved out and a similar number of Blacks and Hispanics moved in.

Like Berwyn, the other incorporated municipality in this study, local politics were characterized by many respondents as being White male-dominated, with rampant cronyism and little opportunity for

The main street of Blue Island has been struggling, but it still has a small town feel.

Blue Island's train depot area is not pedestrian oriented. Plans are underway to turn the stop into a mixed-use 'TOD' (transit-oriented development).

grass-roots or minority impact. Blue Island has struggled with decline for decades (a librarian in Blue Island described it as having 'low self-esteem'), but many are hopeful that times are changing due to the recent efforts to stimulate redevelopment around the regional rail line.

Berwyn borders Chicago on the west. It was originally known as a Czech and Bohemian neighborhood, with many residents working in the nearby GE plant. Large, older parks with Czech names, like Proksa Park, testify to this ethnic heritage. During the last couple of decades, however, there has been a dramatic demographic shift, with a large influx of Hispanic residents. The population shifted from 7.8 to 38% Hispanic in the span of one decade (1990–2000). According to one school administrator, the school population of Berwyn is 75% Hispanic.

Berwyn is well served by major transportation (a point made by virtually every Berwyn respondent). It is traversed by the Eisenhower expressway, the Metra regional railway, and the city of Chicago's rapid transit system (the 'EL'). It is a major part of Chicago's 'bungalow belt' and the rich architectural details of the bungalows are a source of great community pride. It is also famous for its streets of apartment buildings – two-flats and three-flats – built, along with the bungalows, in the 1920s through the 1940s.

Business owners in Berwyn say that the 'number one weakness' of the community is its inability to attract more business. The business community is still feeling the pinch from the loss of Penney's and Sears to an outlying mall (North Riverside). Some remarked that the economic boom of the 1990s seemed to have skipped over Berwyn. There were often referrals to the political in-fighting of Berwyn, which seems to have thwarted any coordinated efforts to spur planning and economic development. A recent election in which a large number of the 'old guard' politicians were replaced by new leaders, some Latino, seemed to offer hope for change. The newly elected mayor called it a 'revolution'.

Berwyn is somewhat divided geographically and socioeconomically along Cermak Avenue, the main commercial core once known as the 'Bohemian Wall Street'. Park and school districts are divided into north and south, which many residents said was the main source of whatever division between north and south existed. The northern part of Berwyn is described as being more low-income and working class, while the southern portion of Berwyn, closer to the train station, is considered more affluent.

Berwyn is well served by rail. The regional rail, Metra, stops at Berwyn and delivers commuters to the Chicago Loop in less than 30 minutes.

THEMES

The value of diversity

The vast majority of respondents claimed that diversity was something they valued, that they were very much aware of it, that diversity promoted tolerance and creativity, and that most people got along fine, despite their differences. Most people reported that getting along with people unlike themselves was simply a reality to be dealt with. A resident of Berwyn put it this way:

> It's easy to be prejudiced and be critical if you don't have to deal with 'them'. But if they're your neighbors and you see them at the 'Y' and you see them at the store and you see them at school, then you better accept it and try not to be prejudiced.

Long-time residents said that the areas they lived in were much more diverse now than when they were children, and although they recognized the challenges this entailed, they did not present it as something negative. There were even claims that the growth some areas were experiencing was in part a result of the lure of diversity – that people actively sought it out.

The degree of tolerance and open-mindedness might have been exaggerated, especially given that many of the respondents were in positions of leadership or were representing community-based organizations, and thus would be strongly inclined to present a positive, inclusive point of view. While many noted that Chicago has long been considered one of the most segregated cities in the US, respondents seemed almost philosophical about the diversity they were living in, saying things like 'diversity is absolutely essential for healthy communities'. They believed that living together was a way to build tolerance. The Alderman of West Ridge, Bernard Stone, made the point that following September 11, 2001, there were riots all over the world, but not in the religiously diverse neighborhood of West Ridge. 'People working together find out that people are really more alike than they are different', said Alderman Stone. This was a common theme amongst the majority of respondents.

Long-time residents noted that attitudes toward race and ethnicity had changed dramatically. This change was especially true for Blacks, some said, who had previously been shut out of places like West Ridge, Irving Park and Portage Park. Some places experienced race riots in the 1970s. A resident of Portage Park said that 'people who grew up here were raised in the 'old school' way – Ukranians lived here and Swedes lived there … If you were from Bridgeport then people said "oh, you must be Irish"'. In other words, awareness of the connection between where someone lived and what nationality or race they were used to be much more pronounced.

Many had the view that people want to find common denominators that unite them cross-culturally. In the West Ridge community, for example, Amie Zander of the West Ridge Chamber of Commerce pointed out that women from the Indian, Pakistani and Orthodox Jewish communities were united in that the women tend to stay at home rather than work outside the home. There were also pragmatic views about unity. The head of one local bank saw the need for everyone getting along as essential because it was 'best for business'.

Despite the generally positive views about diversity, respondents were also able to offer insights about the challenges it presented. Many seemed resigned to the fact that there would always be tensions between groups. Residents of Berwyn expressed fear that the sudden rise in Hispanic population was 'taxing the system', as large families required services like schools, and, put additional stress on utilities and police and fire services. One resident said that there was a strong sense of community in Blue Island but it did not come automatically: 'we work very hard at it'. Another noted that the difference in child-rearing practices between apartment dwellers and single-family homeowners created a clash in parenting styles: 'the apartment kids are given more leeway because there is less space inside the home'.

There were recurrent views about what was keeping these places diverse. Many believed that transportation was the major factor, and all of the communities were identified by residents as having excellent access to the downtown. A resident of Blue Island made the point that city services were key to retaining the affluent end of the diversity spectrum, asking: 'Who's going to buy the more expensive homes we have here if the streets aren't swept and the trash picked up?'. Another said that a mix of housing ages helped racial diversity because 'Whites like the older homes and Blacks like the newer housing'. Others noted that small 'mom and pop stores' that could be walked to were essential for retaining lower-income people.

Some respondents identified rental housing as the key to sustaining diversity. They viewed the rapid conversion of apartments to condos as working against that. This point of view contrasted with political leaders who favored ownership because it results in more tax revenue. But residents associated with community-based institutions noted that rentals are particularly important for new immigrants. One respondent pointed out that if entry-level home ownership is later available to immigrants within the same neighborhood, it provides a degree of stability, because residents can become homeowners without having to move out of the neighborhood.

Development pressures

Housing cost was identified as a key barrier to sustaining diversity. There was widespread recognition among the respondents that new condominium development was putting a stress on the ability of neighborhoods to stay diverse. However, none of the six diverse areas was experiencing the kind of rapid gentrification seen in places like Wicker Park and Buckstown on the North Side of Chicago. As one respondent put it, 'taxes are going up, but it isn't to the point of panic'. (Note that most of the diverse neighborhoods studied here are protected by new tax rules that put a cap on property tax increases. This was not the case when Wicker Park and other neighborhoods closer to the Loop were rapidly gentrifying.) Several respondents in Bridgeport argued that new condominiums were being bought by 'old-timers' who wanted to cash in their equity on their single-family home and move into something smaller.

Concerns about affordability were contrasted with desires for economic growth and new investment. In Berwyn and Blue Island, a number of respondents were more concerned about stimulating economic expansion than maintaining affordability. A respondent in Berwyn, for example, said: 'We need to raise the median income in Berwyn; we're becoming a lopsided town, we're becoming one class'. This sentiment was connected to a concern about ethnic immigration.

Some residents believed that their neighborhoods exhibited a certain robustness – that their communities could withstand change and new development because they 'had seen it all before'. By the same token, respondents had little conception of what government or planners or anyone could do to help support and sustain diversity. They tended to view the existence of diversity as a historically derived, somewhat random occurrence. In addition, rising costs tended to be associated with new condo development, not downzoning or other barriers to development put in place by residents. One resident of Old Irving Park, which had recently downzoned its neighborhood, reasoned this way about the inability of non-affluent people to move into wealthier Old Irving Park: 'it's not about people's attitudes, it's just that housing cost is a barrier'.

Residents were also aware that diverse neighborhoods could become the victims of their own success – i.e., that people moving to a place because of its diversity would ironically end up making the area less diverse. They believed that new condominium development and million-dollar homes on gated cul-de-sacs were not well integrated and could put stress on the long-term stability of diversity in the neighborhood. Some believed that minorities were being pushed out. This perspective varied, however. One developer argued that since new single-family development in these highly built-up neighborhoods involved reclamation of industrial and under-utilized commercial land, developers were making better use of land and not simply displacing existing families: 'Nobody's being pushed out. Nobody's being priced out', a developer in Bridgeport claimed. Yet some recognized a connection between maintaining an industrial base and maintaining diversity.

Many residents believed that a lack of affordability was driving long-term residents away. They argued that it was not about crime, not about traffic or other aspects of urban living, but rather the high cost of living. They also commented that residents who moved further out to gain affordability still retained a connection to the neighborhood, coming back every Sunday, for example, to attend church. Some noticed that this practice was putting a strain on the neighborhood because of the need to retain large parking lots for people to drive in from the suburbs – parking lots that sat vacant most of the week.

There was widespread belief that accommodating neighborhood change – from garage conversions to new 'cinder block condos' was destroying the character of the neighborhood. With little planning or design review in place, changes were perceived as being mostly detrimental, where the correct and only response would be to block change. This sentiment prompted residents of Old Irving Park to fight density and successfully downzone their neighborhood. They pointed to neighboring areas like Wicker Park, where 'there's architectural sameness ... they're all starter condos, with a little balcony for the Webber grill'. Many residents valued the pedestrian orientation and walkability of traditional neighborhood design, and wanted to keep the basic structure – minimal setbacks, wide sidewalks, accessible storefronts – intact.

Residents seemed willing to support density increases along commercial corridors, but not in residential areas. One resident of Old Irving Park understood the advantage: 'We want to get enough density to support a Starbucks'. A city planner assigned to Portage Park said that residents there were 'fine with chains' – they just wanted their retail district to be strong. From the small business-owner's point of view, however, high end chains would have to co-exist with independently owned stores. One business owner said: 'The ethnic stores don't want an "American look"; they will lose business if they look too American.' In a similar vein, a business owner in Blue Island equated the sprucing up instigated by the Main Street program as 'harassment', constituting 'stupid, picky little rules that will push people out'. On the other hand, the idea of making a business area 'attractive' was seen as significantly less important than simply attracting a major retailer to the area: 'A Walmart or a Kmart — *that* would help diversity', said one Blue Island business owner.

There was widespread resentment about density increases occurring in haphazard ways. One issue that was discussed repeatedly was the practice of converting single-family houses to multiple-family residences. As one respondent put it: 'I don't care what ethnicity they are, if they're cramming into one house it causes problems'. Adding illegal apartments, motivated by the need for additional rent, was cited as a significant problem in Berwyn. Some pointed out that the poor design of apartment buildings could make multi-family housing more of a liability than it needed to be. There was some sense that lack of planning and design for development was putting a strain on things. 'That's one of the problems with this town', a respondent said of Blue Island, 'there isn't really a design to it'.

Social connectedness

While recognizing that people are now more tolerant of diversity, many respondents believed that 'better networking' was missing in their communities. Some believed there was a lack of ability to work together to address common problems, such as the ability of businesses to band together to address parking and litter problems. One business owner in Portage Park said that it was hard to get ethnic groups involved in business organizations because 'they don't think in terms of the neighborhood as being their market, they think in terms of their group being their market'. Latinos in Berwyn were described by several residents as 'insular'. The head of an historical society in Blue Island said that he had difficulty getting Hispanics involved in his organization because of a lack of historical connection: 'I have no photos, artifacts ... nothing from the Hispanic community'. There was some sentiment that people in diverse places were having a difficult time forming a unified vision of what their neighborhoods should be.

Many contended that an increase in mobility was severing connections to the local community. They pointed to the increased role of the automobile in this. A resident of Berwyn recalled when police and fire-fighters were required to live in the communities they served, so that they would have a stake in the local area. Some recalled a time when they knew where all of their teachers lived. Others noted that there were few events that brought people together. There was a pervasive view that involvement in various neighborhood-based organizations and institutions had declined dramatically. A fireman in Blue Island told of how the Lions and VFW Hall were gone, and that 'a lot of fraternal organizations are going by the wayside'.

Some respondents said that people only care about their own private worlds, and no longer know how to function as neighbors in a neighborhood. One long-term resident of Portage Park recalled: 'Growing up, nobody had a tall fence. Now everyone has a 6 foot tall wood fence and everybody just drives into their little piece of the world'. Sometimes this was attributed to newcomers. In Bridgeport, one resident said that when new townhomes get built, 'First thing that goes up is the outside security door, then a security fence that locks it. People in Bridgeport laugh at it because there's no need for it – "it's not that bad, folks!"'. For some, the implication of this was a loss of caring about how to be a good neighbor. One argued that 'some people don't understand how a neighborhood works, how to keep it on a social level; they just don't understand what has to be done, like cutting the grass'.

But there was also an interesting connection drawn between tight-knit, closely bonded communities and decreased mobility. In West Ridge, the tight bond of certain ethnic and religious groups was believed to be tied to the ability to walk to communal facilities like synagogues. The Orthodox Jews of West Ridge are one example. Diverse communities were identified as providing access for new immigrants – to services and facilities as well as people who speak their own language. Some respondents said that ethnic immigrants created permanence in the community because they were so attached to their neighborhoods. The strong bonds within different groups were also cited as a reason why gentrification pressures did not always amount to much. Though imposing some hardship, respondents believed that residents would find a way to stay in the community because of the cultural organizations and institutions they had come to depend on.

The role of public and private institutions

Most respondents believed that the role of public and private institutions in community life had changed significantly. While religious institutions were still thought to be the center of social activity in some neighborhoods like West Ridge, respondents in Bridgeport saw a significant decline in the role of church and school. Catholic parishes in particular had gone from being the center of social and educational life to being of much more limited social significance. This was attributed to a broadening of resident interests and lessening of group affiliation. Some argued that new residents were not joining parishes and not sending their kids to local schools: 'the new people are not joiners; there isn't a sense of community with these people'.

There is a certain paradox to having strong institutions in diverse neighborhoods. Such institutions had historically tended to separate people by enforcing attendance at a particular parish and school, with little outside contact. Few facilities, functioned specifically as inclusive community centers. Now, religious, ethnic and various private institutions are seen as being crucial for maintaining some degree of social connectedness. The Irish-American Heritage Center in Irving Park, for example, is run by a board that includes 'lawyers and doctors in addition to blue collar and people of little means'. Arts organizations like the Portage Park Arts center were seen as drawing very diverse groups together: 'Housewives doing quilting next to people doing eastern meditation exercises'. An artist in Bridgeport claimed that the arts district was 'holding Bridgeport together'.

Many believed that more community facilities were needed. A resident of Irving Park observed that communal play spaces were essential not just for keeping people connected, but for taking the kids off the streets and giving them somewhere to go. This in turn, she argued, could lessen tensions among neighbors. In a neighborhood with high density and insufficient public space, the provision of spaces for play may be essential for keeping neighbors neighborly. Some observed that there was a need not only for more facilities, but for better access to existing ones. One resident of Blue Island said that children had poor access to public libraries, and many had to cross over railroad tracks to get there. A similar comment was made about access to Berwyn's library.

Some believed that the fragmentation and privatization of communal places meant that some groups lacked social footing. The director of the West Ridge Chamber of Commerce said that there were few places for teenagers to go, for example. It was claimed that the Blue Island community lacked sufficient 'lifestyle options' for seniors. Places that would have functioned as community focal points in previous decades no longer existed. This includes centralized commercial areas. In Portage

Park, the six corners shopping district was said to be the central life of the community, and its steady decline bothered many Portage Park respondents.

Some believed there had been a breakdown in their community's institutional strength. They noted that under-funding of libraries and even the lack of a local newspaper was making it more difficult to hold the diverse community together. They pointed out that it used to be that Chicago public schools played a formative role in community life, but that funding cuts had eliminated the ability to keep the schools open to the public after hours. Respondents also noted that public schools were often not servicing local children, and many parents, particularly newer affluent ones, opted to send children outside of the neighborhood to private schools. In Irving Park, for example, the landmark Carl Schurz high school was populated by students being bussed in because 'kids from the neighborhood try to avoid going there'. Several interviewees were aware of recent gang-related shootings at the school, and attributed this to outsiders.

DISCUSSION

Socially diverse neighborhoods are distinguishable as places where residents are working out the day to day complexities that arise when different kinds of people occupy the same geographic space. These neighborhoods are not gentrified – although they often have strong gentrification pressures – and they are not suffering from high crime and disinvestment. But neither are their business districts and schools thriving. These are neighborhoods and communities that are sometimes viewed as stops along the way to the ultimate American dream, the more affluent suburban neighborhood. As the mayor of Berwyn summarized:

> We've always been a stepping stone community, a launching field for the Napervilles ... Berwyn started out with the Czechs, then the Irish, then the Italians, now the Hispanics; they come from Chicago. Some stay here and some end up moving further out. It's the next step in the economic ladder.

There is a constant tension arising from paradoxical change: development pressures fighting against neighborhood preservation efforts, older public facilities that deteriorate despite new investment, small businesses that are pushed out by chain stores, local institutions with far-flung constituencies, declining schools in the midst of rising wealth, old-timers that are displaced by young professionals, and real and imagined fears that gradually escalate. Eventually, such neighborhoods often veer in one direction or another, and lose their diversity.

Our respondents seemed to have a clear understanding of these issues and tensions. They were very much aware of the diversity surrounding them; diversity had meaning. They were also aware that urban investment and revitalization activities had an effect on diversity. Investment can be good for social diversity in one context, and work against it in another. Overall, the interviews revealed that the struggles between different groups were interwoven – that is, expressed through and around – issues very pertinent and even central to the design of the built environment.

Yet all six communities lack basic planning and design support. Since development tends to be smaller in these already built-up neighborhoods, neighborhood plans and designs are the only mechanism for ensuring that development fits into a broader framework. Unfortunately, most neighborhoods in Chicago do not have any kind of legal power to enforce a collectively derived vision. None of the six communities have an open, democratically driven neighborhood planning process. The process is closed, mostly a matter of the personal interest of individual aldermen. As a developer from Bridgeport summarized it, 'In Chicago, the aldermen control development because the aldermen control zoning'. Even for large-scale projects, there is generally no neighborhood involvement or community review of projects. When Alderman Stone was asked why a large park was being developed near his home at the suburban edge of West Ridge, he said, 'Well, I like it … I have a personal interest in parks'.

The lack of public involvement in diverse places is unfortunate, because in these places especially there is a need to foster public participation and collective input. That's because diverse areas are likely to be located in older, inner-ring suburbs that are mostly built out, and prone to recurring issues over the appropriate use of space – space for parking, for new development, for single-family conversions, for public facilities, for schools. Our interviews confirmed that the ever-present contestations over issues endemic to older neighborhoods had no process of resolution – no participatory approach that would ensure an open attempt to strike a compromise or a consensus.

The practical value of planning and design is that it exerts pressure on local aldermen to pursue a collective vision and not deviate according to their particular whims or, many respondents believed, their ability to line their own pockets. Aldermen respond to their constituencies, but, many believed, only to those with the loudest voices. There was an example in Bridgeport of a developer who was trying to build town-homes in a single-family area that would have included common open space, but the alderman insisted that only single-family housing could be built. Without consideration of housing mix, context, need, design and planning the alderman was only using the blunt instrument of conformity to respond to the development proposal.

Lack of planning impedes the community's ability to feel like it's moving forward and progressing on some level. Many respondents had the view that certain neighborhood developments or improvements were essential to the neighborhood's long-term health, and that without them, the neighborhood would decline. With so much riding on certain high-profile projects, much could be gained by concretizing these developments as essential parts of community-generated plans.

In the absence of a collective vision, residents in a diverse neighborhood may be particularly prone to antagonism toward new development. Without clear plans and guidelines, change is often seen as threatening. There were many examples given by respondents of attempts to block new development – especially development that would increase density. To the degree that some new development is necessary to sustain diversity, neighborhood planning and design could play a stronger role in ensuring compatibility of diverse types of development, thereby minimizing attempts to block changes that would actually promote diversity. Residents seemed to be most concerned with predictability and the ability to avoid bad quality design, not with blocking people out per se. Residents in Berwyn were upset about the loss of historic structures to make way for an Applebees, and a parking garage proposed for Devon Avenue in West Ridge was widely criticized by neighborhood residents and merchants because of its bad location and design. These types of stresses may have been dealt with in a less contentious way if there had been a different kind of decision-making environment.

The failure of process also tends to result in uncreative solutions to things. The city planner for Portage Park said that some residents believed that the way to battle crime was to promote higher housing prices. Perhaps if there had been a process in place for addressing issues like neighborhood crime, solutions that seek a more holistic understanding of proposals, especially their negative consequences (like gentrification and residential displacement), would be a matter of course. Tensions over practical issues like traffic and parking should not be allowed to escalate and transfer into objections about particular groups of people. There is a need to empower residents with the tools and strategies they need to address such issues. Providing design ideas that accommodate mix, foster connection, and provide security may be one essential step in the right direction.

NOTE

1 Much of this material was drawn from the interviewees themselves, with cross-referencing from *The Encyclopedia of Chicago* (Grossman et al., 2004).

PART THREE

THE STRATEGIES

I have been arguing that the promotion of social diversity has implications for the design of the built environment – that the 'container' should exhibit some relation to its 'contents', as Lewis Mumford (1968a, p. 165) put it. Above all, social diversity requires an environment that is designed for different types of people – people with different needs, interests, tastes, and constraints. This mix, in turn, requires special attention to how people are connected, and whether they feel secure. Design for diversity is largely a matter of being strategic about where development occurs, and what form that development takes.

The strategies and design proposals given in the next three chapters tend to emulate the physical conditions of the pedestrian-oriented city, an urban form based not on cars, but people. This does not mean that the social and economic conditions of the car-based city are being assumed away. But it does mean that it is now necessary to design a socially mixed city in an environment largely devoid of the mechanisms that had previously kept social separation in place, despite geographical proximities – the rules of deference as discussed in Chapter 2. The strategies for place diversity presented here are in that sense radical – they attempt to do away with a class-based social geography.

It is useful to look briefly at the motivations people have had for designing the built environment in various ways. Design may be intended to benefit specific populations, such as women (*Design and Feminism*, Rothschild and Cheng, 1999), children (*Creating Child Friendly Cities*, Gleeson and Sipe, 2006), or the elderly and disabled (*Inclusive Urban Design*, Burton and Mitchell, 2006). Design proposals may be most concerned with environmental conservation, civic vision, or social interaction (Barnett, 2006). Or design might be particularly attuned to the requirements of different kinds of users, helping them navigate a space, feel safe, be socially active, or find spiritual fulfillment (Greed and Roberts, 1998; Montgomery, 1998).

Most often designers of urban places are focused on creating a 'sense of place', as in the work of the Project for Public Spaces (2007). Or they may be hoping to change people's behavior, as in strategies for encouraging the use of public open space (Rishbeth, 2001). Recently, design has been used in an effort to promote public health (Moudon, 2005). These various strands represent all kinds of different goals – socialization, excitement, profit, crime reduction, civic pride, health. Evaluative criteria that correspond to these goals are based on principles like human scale, pedestrian access, safety, complexity, legibility, or spatial enclosure.

Urban design theory has been used as a way of interpreting the diversity of urban places. There are those who can look at sprawling landscapes and see the basis of diversity, as implied in Soja's declaration that the 'representations of spatiality and historicity' in Los Angeles are 'archetypes of vividness, simultaneity, and interconnection' (Soja, 1989). There are also esthetic explanations of fragmentation and difference (largely devoid of explicit social meaning, or highly abstracted from it), as in Rowe and Koetter's (1978) *Collage City*, or Tschumi's (1996) *Architecture and Disjunction*. Theory may be used to better understand the symbolic content and underlying meaning of designed spaces, or to invent new vocabularies (Soja, 2000). In these contexts, normative rules about the urban experience found in texts such as Alexander *et al.*'s (1977) *Pattern Language* or Lynch's (1981) *Good City Form* may be downgraded as 'trivial operational features of urban environments' (Cuthbert, 2006, p. 12) and a far cry from 'theory'.

Finding a base of support for strategies that uphold social diversity may be less about finding an appropriate theory and more about understanding the implications of specific social goals. In fact virtually every consideration in the design of human settlement can be thought of as having some sort of social value, either implicit or explicit. The task, then, is to figure out what social goals associated with various planning and design approaches can be linked *specifically* to social diversity. To get ideas about this there are a number of design perspectives to draw from.

Design for equality. Design based on the idea of equality could potentially have a lot to do with fostering diversity. For example, if design for equality is about equalizing access and ensuring closer proximities between where people of all ages and both genders live and work, as exemplified in Dolores Hayden's (1980) proposal for a 'non-sexist city', then the idea of equality becomes very much intertwined with the notion of supporting diversity. Often, in fact, design for specific constituencies is motivated by a desire to equalize access for all groups and therefore make up for past inequities.

Design for imageability and vitality. The primary goal of most urban design is to create a 'successful' human environment, which usually means an enhancement of the visual, aesthetic experience. The focus might be to provide some visual coherence, some framework for making sense of the urban realm. Kevin Lynch's (1980) 'dimensions' were aimed at achieving successful urban places, and what constituted success was largely a matter of performance criteria, such as imageability and legibility. Lynch wanted the built environment to respond well to people's needs, to allow people to be successful, to give people access to things, and to provide vitality.

In this mode is the idea of 'place-making'. Place-making, in turn, is largely predicated on vitality, activity, and liveliness. Whether viewed as the image of the city (Lynch, 1960), the city as a set of sequences (Cullen, 1961), or the idea of instilling 'permanence, locus, individuality and memory' (Gosling, 2003, p. 22), these dimensions are about creating a successful experience for the individual urban resident. That experience is successful if it promotes exchange, generates lots of activity, or presents a positive, culturally meaningful experience (Montgomery, 1998).

Diverse places, too, need to be 'successful' and have a sense of place. The issue, as discussed in Chapter 3, is how to balance imageability and vitality with the other requirements of diverse places.

Design for organic wholeness. Some designers are more concerned with the generative processes that create successful environments, as opposed to prescriptions for end-state forms. The aim of Christopher Alexander, for example, is to discover the laws that produce an organic wholeness – cities with 'organic, personal, and human character' (Alexander et al., 1987, p. 234). That organic quality defines their success.

It could be argued that organic growth – wholeness – is necessary for social diversity too. According to Alexander's theory, growth should be piecemeal, and every increment of building should contribute to a larger whole. Rules for incremental change are likely to be supporting of diversity because they accommodate finely grained elements, as opposed to singularly conceived, monolithic elements that are likely to contribute to homogeneity.

Design for community. Sometimes the design of cities is called upon to provide 'a physically humane setting for a social existence'

(Gosling, 2003, p. 7). The emphasis is on existing neighborhoods and the way public spaces can be designed and located to strengthen existing or new communities. Morrish and Brown (2000) identified five types of neighborhood assets – homes and gardens, community streets, neighborhood niches, anchoring institutions, public gardens – that should form the basis of community-oriented planning and design.

Often design is directed at the need to foster social interaction. The New Urbanists, for example, believe that the design of neighborhoods should adhere to certain principles in order to satisfy the ideal of community. Neighborhoods should be designed to be walkable, centered, and perhaps bounded; their outdoor spaces should be enclosed, traffic should be calmed, and public space should be plentiful and evoke civic pride. Such design principles are intended to enhance civic awareness, and therefore community.

The connection to diversity is not straightforward. A key paradox confronting attempts to 'build' community through physically-oriented policies is that, at least at the neighborhood level, such community building efforts have been historically linked to efforts to promote social homogeneity and exclusion (Silver, 1985). David Harvey likened the quest for community via design to 'surveillance bordering on overt social repression' (Harvey, 1997, p. 69). Yet, the effort to design for social connection, the maintenance of social inclusion, and feelings of security within a place are goals that are essential for both community and diversity.

Design for sustainability. Finally, there is the connection to sustainability. Ian McHarg, in *Design With Nature* (1969), was an early advocate of the idea that urban design should be based on ecological principles. Since then, calls by ecologically-oriented planners to be responsive to the natural environment and to foster diversity in human realms have been translated to certain design principles. Calthorpe (1993), for example, defines his work as one of attempting to give specific form to sustainability goals like diversity and interdependence (Calthorpe, 1993). In an article titled 'Urban design to reduce automobile dependence', Newman and Kenworthy (2006) show how features like pedestrian sheds and town centers affect sustainability. Elements of sustainable city form include containment, public transport, access to services, housing variety, self-sufficiency, adaptability, and local autonomy.

THREE DESIGN STRATEGIES

All of these design motivations – equality, imageability, vitality, organic wholeness, community, sustainability – are good sources for ideas about designing for diversity. They help to envision the physical conditions that correlate with, promote or sustain diversity. To make sense of these varied ideas, and to hone in on those likely to be most relevant to social diversity, I propose a framework that consists of three inter-related strategies – mix, connection and security.

Design strategies for diverse places involve reading the environment in a particular way. A street and a sidewalk become places not just to accommodate movement, but places that might have special significance for the support of social diversity. Strategies are also largely about prioritizing and making strategic investments. Where can a relatively small amount of attention go a long way in promoting and sustaining the social diversity that exists? Can we identify priority areas that become the focus of stabilization, and what kind of intervention would be appropriate?

The next three chapters outline the basic rationale for these three primary strategies – mix, connection, and security. Implementation of each involves either development, preservation, or mitigation. That is, supporting mix, connection and security involves either interjecting design elements that support diversity, finding elements that support diversity and preserving them, or finding what is detracting from diversity and trying to lessen the negative effect. There is a strong emphasis on analysis – how to evaluate the built environment for its diversity-sustaining potential.

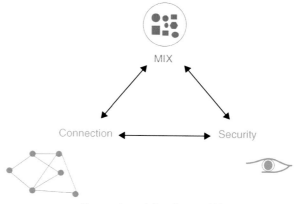

Three requirements for a diverse neighborhood

7 Mix

There are two types of strategies related to mix in diverse places: the mix of housing options, and the mix of services and facilities needed to support a diverse set of needs.

HOUSING MIX

Housing mix is needed to ensure that social mobility does not require geographic mobility – in other words, providing opportunities for residents to change their housing 'in place'. Housing mix steers us away from the idea that neighborhoods represent monocultural reflections of social standing, stepping stones 'in a trajectory of ascendance through a geography of status and income' (Espino, 2001, p. 6). Diverse neighborhoods are those that provide multiple opportunities.

Three things can help sustain a mix of housing options:

- Mix of housing types – different tenures (owner vs. renter occupied), and different forms and sizes, from single-family to multi-family.
- Mix of housing ages – the retention of existing housing stock, integrated and blended with new housing stock. This is because older units are often more affordable than new. Diverse neighborhoods tend to have historic layers, which can be used to understand the unique qualities of places (Hayden, 2003).
- Policies that sustain a mix of affordability levels.

The policy side of the equation is essential, because the market is unlikely to support a mix of unit affordability levels on its own. The support of a diverse neighborhood requires that one type of housing or price level not be allowed to 'take over'. Most often, there is a need for policies that keep units affordable (this is discussed in more depth in Chapter 10).

One of the most entrenched truisms about city planning is that well designed places quickly become unaffordable. It's a relatively simple matter of short supply and high demand, coupled with the fact that

Bridge Port

■ High form diversity

■ High social diversity

West Ridge

☐ one-story homes and bungalows
☐ two-story homes
☐ small multi-family like duplexes
☐ larger multi-family and apartments
☐ commercial buildings

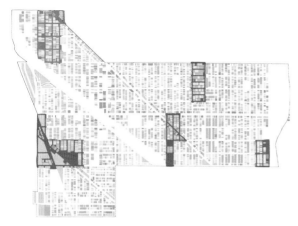

Irving Park   Portage Park

Designers should become familiar with the patterns of diversity in diverse neighborhoods. These maps show blocks with high resident diversity (in purple) and high form diversity (e.g., unit types and sizes, highlighted in red). Also shown is land use by parcel.

affordability in desirable places goes against the basic laws of land use in the American real estate market.

Design strategies that support housing mix include codes that can address housing mix, small unit integration, housing near transit, housing in commercial areas, and new/innovative housing types.

CODES THAT CAN HANDLE HOUSING MIX

Diverse neighborhoods need codes that specifically address issues related to the mixing of housing types. Diverse areas are prone to wide fluctuations in housing size, type and style. This is potentially good for diversity but potentially bad for long-term stability and neighborhood cohesiveness. Codes are needed that go beyond simple FARs and unit sizes, and instead allow flexibility within a framework that promotes a successful integration of housing types.

Diverse areas need form-based codes to accomplish this. The examples shown here are from the 'SmartCode' (http://www.smartcodecentral.org/). A particular zone is regulated in function of its character, not on the basis of use (single-family vs. multi-family).

It regulates height by no. of stories, disposition (where the building is placed on the lot), outbuildings, and parking. Use diversity is encouraged, and buildings can function as residential, lodging, office or retail spaces,

SMARTCODE

municipality

ARTICLE 5. BUILDING SCALE PLANS

SECTION 5.4.11   **T4**

(see table1)

BUILDING FUNCTION (see tables 10&11)

a. Residential	Limited use
b. Lodging	Limited use
c. Office	Limited use
d. Retail	Limited use

BUILDING HEIGHT (see table 8)

a. Principal building	4 stories max, 2min
b. Outbuilding	2 stories max.

LOT OCCUPATION

a. Lot width	18 ft.min 96 ft max
b. Lot coverage	70% max

BUILDING TYPE (see table 9)

a. Edgeyard	Permitted
b. Sideyard	Permitted
c. Rearyard	Permitted
d. Countyard	Permitted

BUILDING DISPOSITION

a. Front setback	6 ft.min. 18 ft. max.
b. Side setback	6 ft.combined min
c. Rear setback	3 ft. min.*
d. Frontage setback	

OUTBUILDING DISPOSITION

a. Front setback	20 ft.min. big. setback
b. Side setback	0 ft. min.or 3 ft.
c. Rear setback	3 ft.* or 23 ft.

PRIVATE FRONTAGES (see table 7)

a. Common lawn	Prohibited
b. Porch & Fence	Prohibited
c. Terrace or L.C.	Prohibited
d. Forecourt	Prohibited
e. Stoop	Prohibited
f. Shopfront&Awnig	Prohibited
g. Gallery	Prohibited
h. Arcade	Prohibited

Refer to summary table 14

PARKING PROVISIONS

See tables 11 & 12

*or 15ft. from center line of alley

BUILDING HEIGHT

1. Building height shall be measured in number of stories excluding a raised basement, or inhabited attic.
2. Each story shall not exceed 14 ft clear, floor to ceiling.
3. Maximum height shall be measured to the eave or roof deck.

BUILDING DIAPOSITION

1. The facades and elevations of principal buildings shall be distanced from the lot lines as shown.
2. Buildings shall have facades along principal frontage lines and elevations along lotlines. (see table 16E).

OUTBUILDING PLACEMENT

1. The elevations of the out buildings shall be distances from the lot lines as shown

PARKING PROVISIONS

1. Uncovered parking spaces may be provided within the 3rd layer as shown in the diagram (see table 16D).
2. Covered parking shall be provided within the 3rd layer as shown in the diagram (see table 16D)
3. Trash containers shall be stored within the 3rd layer.

SC55

Sections from the SmartCode (above and page 119) show how form-based codes differ from conventional zoning codes. They regulate form – building frontages, height, disposition, placement – rather than focusing on details about the kinds of uses and activities allowed.

within some limits. But the main point is that regulations on form are parametric, allowing a range of options – essential for mixing housing types successfully. Flexibility is allowed within some bounds to ensure coherence.

a. THOROUGHFARE & FRONTAGES

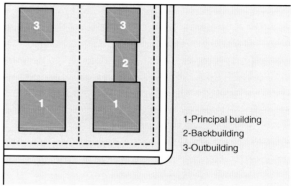

Building	Private frontage	Public frontage	Vehicular lanes	Public frontage	Private frontage	Building

Private lot	Thoroughfare (R.O.W.)	Private lot

c. BUILDING DISPOSITION

1-Principal building
2-Backbuilding
3-Outbuilding

Housing type incongruities, a common occurrence in diverse neighborhoods, could be addressed by applying form-based codes.

Where to focus coding effort

Blocks that have high social diversity and high building form diversity should be identified (areas in blue, below and page 121). These are the locations in which to ensure that codes don't undermine the mix, but instead help stabilize and procure it. This may require the adoption of new, formbased codes (like the SmartCode).

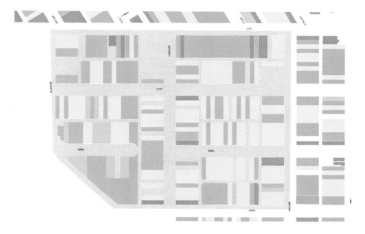

Blocks in Bridgeport with high social/form diversity.

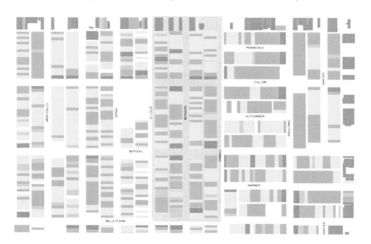

Blocks in Irving Park with high social/form diversity.

These maps show areas in Bridgeport, Irving Park and West Ridge that have both high form mix and high social diversity. These areas should be targeted for coding that preserves mix. New development in these areas should respect existing forms, which is also accomplished through the use of a form-based code.

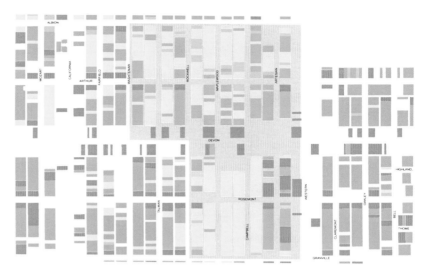

Blocks in West Ridge with high social/form diversity.

High social/form diversity

one-story homes and bungalows
two-story homes
small multi-family like duplexes
larger multi-family and apartments
commercial buildings

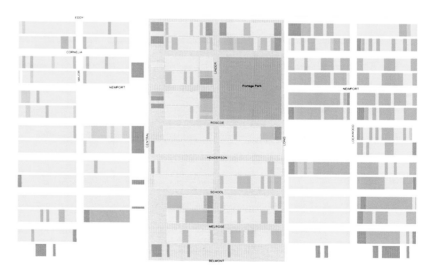

Blocks in Portage Park that lie between areas of one housing type vs. another.

If there are no areas with high social diversity and high form diversity, then look for areas of high social diversity that are near parks or other amenities, and that also seem to have a range of housing types. This is the case with the Portage Park example. The blue highlighted area surrounds a park between two relatively homogeneous housing areas.

MULTI-FAMILY UNITS IN SINGLE-FAMILY BLOCKS

Form-based codes will help ensure that new housing is compatible with existing neighborhood character. New multi-family housing should either have the basic typology of existing multi-family housing (if there are good examples in the area), OR, it could take on the typology of single-family housing in the area (i.e., multi-family buildings with a single-family housing character or compatibility). Examples of the latter are shown below.

Housing type mix that could be more successfully blended through the use of form-based codes. The images are from the West Ridge neighborhood.

Examples of multi-family buildings with a single-family compatibility.

LINKAGES BETWEEN DIVERSE HOUSING TYPES

There should be a concerted effort to create neighborhood link-ages between variegated housing forms. Often in the transition areas between different housing types there is vacant land or under-utilized space. These spaces tend to insulate rather than connect.

a)

c)

b)

d)

Social diversity
☐ low
☐
☐
▨
■ high

Different levels of diversity in a) Bridgeport; b) Irving Park; c) Portage Park; and d) West Ridge. Transitional areas – the places where high diversity adjoins areas that are less diverse – are circled in red.

Finding transitional areas

The first step is to find those places that are harboring a mix of different housing types and focus on them. There are different strategies. For example, a designer could focus strategically on:

- Mixed areas that are isolated – areas of high mix surrounded by areas of low mix
- High contrast areas – places that go from high mix to low mix abruptly, especially those that are on either side of a major transportation artery (see Irving Park example).
- Open space areas – areas with a lot of transitional mix (from high to low diversity) that also have available open spaces (Portage Park and Bridgeport examples).
- Variegated areas – places with very different housing types close together. Open spaces in these areas should be targeted for context sensitive development, preferably based on a form-based code (West Ridge example).

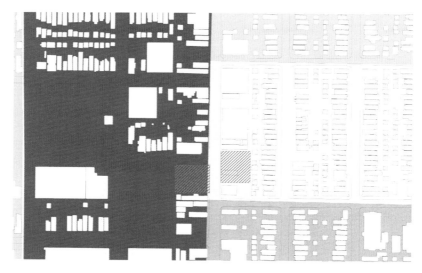

Strategic open areas in Bridgeport.

Designing transitional areas

Because of the diversity of building types and uses, sometimes there are juxtapositions that are quite jarring. This can be exciting – a lake next to a high-rise building is an abrupt transition that can be very vital. But other times the juxtaposition is not nearly so interesting, and may even be disorienting and disruptive.

Strategic open areas in Bridgeport

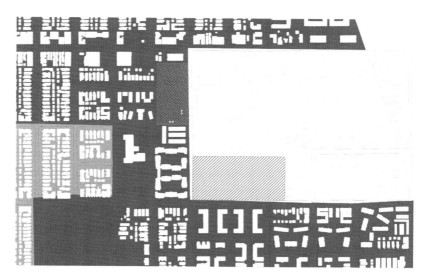

Variegated areas in West Ridge.

In diverse places, there are often abrupt mixes of housing types. The mix can be awkward, with no transition space and no sense of context. The contrast seems only fragmented and ill-planned. It accentuates difference, but not necessarily in a positive way.

Design for transitional areas should focus on creating a physical or perceptual linkage between the diverse housing types that surround it. For example, the development of public space can be a good way to

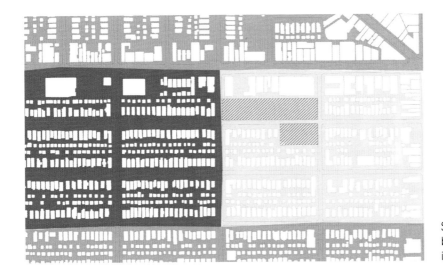

Strategic open areas
behind commercial strip
in Portage Park.

link housing types that otherwise lack any deliberate, meaningful integration in terms of their physical design. If the diverse housing types front the public space equally, they are equally acknowledged as important at least on a symbolic level. They share a collective space, and jointly connect to it. In addition, encouraging buildings that can serve a variety of uses is important for lending vitality to the adjacent public space. This can also help provide a better linkage between the diverse residential types.

Transitional areas to be designed are either 1) open spaces or 2) streets. For open space, parks are an obvious choice. If there are streets

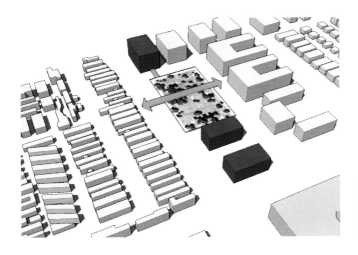

For areas with very different building types, transition space is needed. It's important to have active uses around the space, not vacant lots or buildings.

rather than open spaces between different housing types, the streets can be designed as connecting spaces. Woonerfs – shared streets designed to be negotiated by cars and people – are an innovation that could be particularly useful in diverse urban neighborhoods.

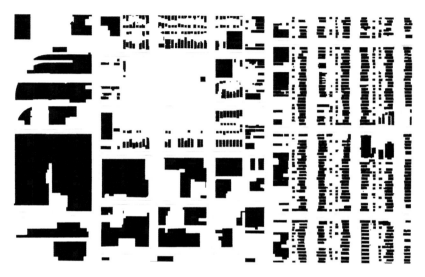

This area of Bridgeport has large apartment buildings and single-family homes in one area. Transitional housing or other developed space is especially needed in the empty lots in between.

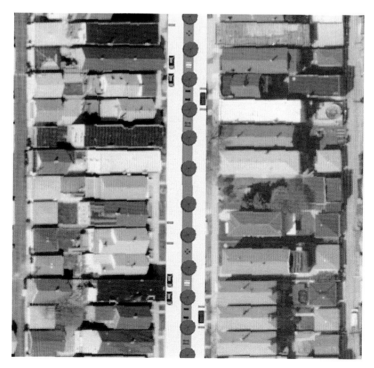

On streets with widely varying housing types or little nearby collective space, consider creating a woonerf (the Dutch term for 'street for living') to help sustain the mix.

SMALL UNIT INTEGRATION

There is a relationship between diversity and density, but it cannot be specified in absolute terms. Even Jane Jacobs recognized that density requirements for generating diversity are likely to vary from one place to another. Densities that are too high could repress rather than stimulate diversity by giving rise to standardization of building type.

The diverse neighborhoods studied here are nowhere near the concentration requirements envisioned by Jane Jacobs, who was advocating something in the range of 100 dwelling units per acre.

The ability to establish urban-level diversity in less intense settings is a challenge, but there are some basic design strategies that could be useful, and that are likely to be essential for sustaining diversity. Small unit integration is especially important because it provides options for low to moderate income households, as well as additional rental income for existing property owners.

Housing near transit

Pressure to increase density can be relieved by supporting small unit integration – accessory units or 'granny flats' are one example. Sometimes residents object to allowing accessory units because of a

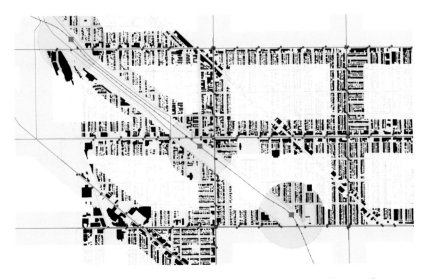

The yellow circled areas in Irving Park, above, are less than a five-minute walk to rail transit. These areas should be targets for infill development.

(real or perceived) increase in cars in the area. Tying the provision of small units to areas well served by transit can lessen this impact, as well as lessen the burden of car ownership for low-income families. New affordable units should be allowed and even encouraged in areas close to transit – with parking restrictions, not requirements.

Housing in commercial areas

Housing over commercial space is another way of promoting housing mix via small unit integration. It is not only smaller and therefore more

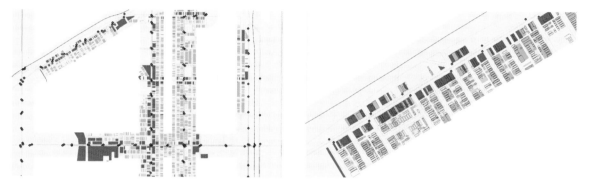

Blue areas are very close to public transit; they also contain varied housing types. Encourage small unit integration in these areas especially.

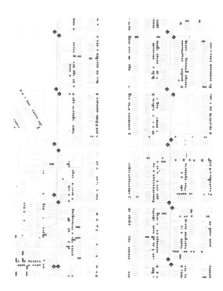

The dark squares are accessory buildings (garages) within 500 ft of a bus stop in this section of Bridgeport.

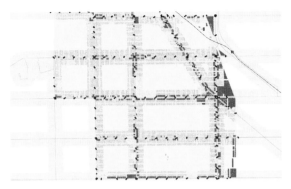

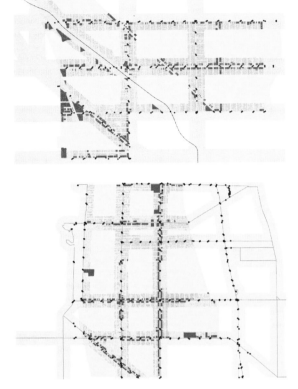

Irving Park, Portage Park and West Ridge have a lot of acreage within a short 500 ft of a bus stop (highlighted in blue). These areas should be targets for infill development – small accessory units in residential areas, and housing over commercial space in the commercial areas.

affordable, but has the added value of providing more client base for small business owners in the vicinity.

INNOVATIVE HOUSING TYPES – COURTYARDS AND CLOSES

Diverse neighborhoods in 21st century America could make use of innovation in multi-family arrangements to better accommodate a variety of housing sizes and types. Courtyards and closes (short looped streets with housing around them) are especially appropriate.

Garden City designers like Raymond Unwin and John Nolen were masterful at integrating housing types. They were especially good at fitting in attached row houses amongst single-family housing. The examples shown here are from Unwin's classic text *Town Planning in Practice*, written in 1909.

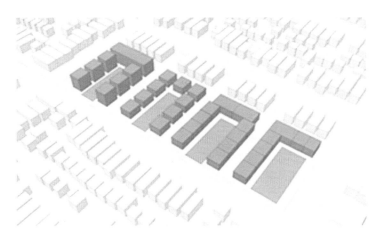

Courtyard housing may be a great way to integrate smaller housing types in a diverse neighborhood. Many examples are shown in the book *Courtyard Housing in Los Angeles: A typological analysis* by Polyzoides *et al.*, 1997.

Housing type integration was important to early 20th century planners. Below are a few examples of their innovative arrangements.

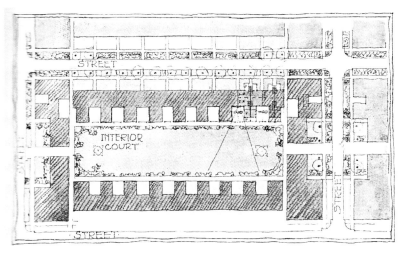

THE ORDINARY CITY BLOCK, EVEN WITH ATTACHED DWELLINGS FOR FIFTY OR SIXTY FAMILIES, CAN HAVE A BEAUTIFUL INTERIOR GARDEN COURT

William Drummond, 1913. From the *City Club of Chicago Quarter Section of Land Competition*. Published in Yeomans, 1916, p. 41.

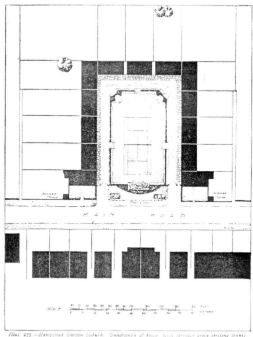

Illus. 277.—Hampstead Garden Suburb. Quadrangle of house with carriage drive circling tennis-court, built for the Hampstead Tenants, Limited. See Illus. 292.

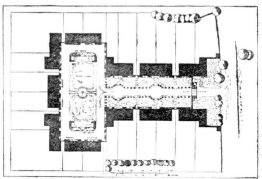

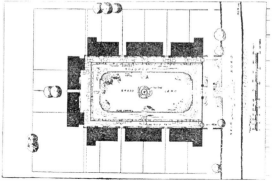

Illus. 278.—Pairs of houses arranged round a green

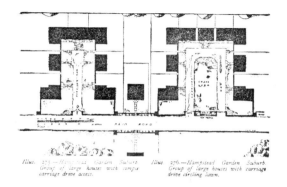

Illus. 275.—Hampstead Garden Suburb. Group of large houses with simple carriage drive access.   *Illus. 276.—Hampstead Garden Suburb. Group of large houses with carriage drive circling lawn.*

Illus. 279.—Arrangement for developing greater depth of land by means of carriage drive.

Raymond Unwin, 1909. *Town Planning in Practice*. These four examples are found on pages 353–355.

Mix lot sizes

Finally, there are certain structural aspects of the form of neighborhoods that contribute positively to mix. One is the diversity of platting configurations, often related to the variation of street types (see Marshall, 2004a). Platting diversity may help promote housing type diversity. Lot configurations may range from uniform on one block and mixed on another, or large, perpendicular lots on one or both ends with smaller lots in between.

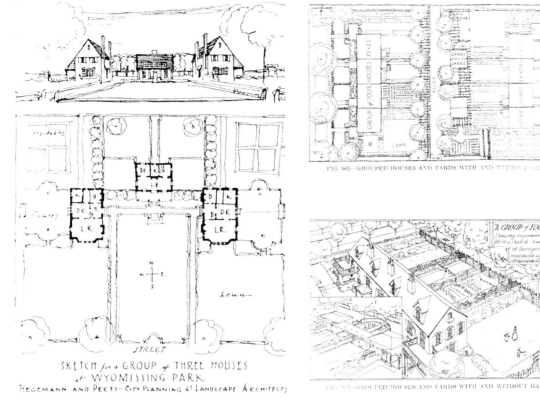

Hegemann and Peets, 1922. From *The American Vitruvius: An Architects Handbook of Civic Art*, p. 213.

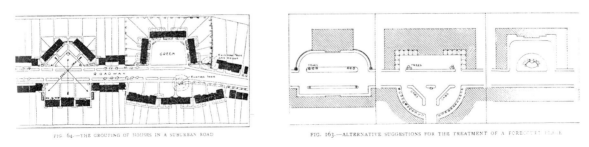

H. Inigo Triggs, 1909. *Town Planning: Past, Present and Possible*, pp. 195 and 315.

In Chicago, the variety of configurations on a given block is no doubt supported by the narrowness of Chicago lots – often 25 by 125 feet in older neighborhoods. But in addition, as Keating (2005) documents, subdividers provided a range of lot shapes and sizes explicitly for the purpose of providing for a range of incomes.

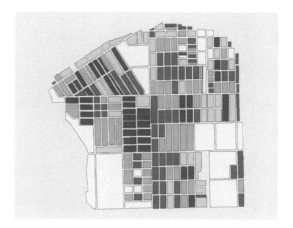

The Bridgeport neighborhood has a good mix of platting configurations – each color shown here corresponds to a different arrangement of lots. Platting diversity promotes housing type diversity.

FACILITIES AND SERVICES MIX

A diversity of people needs a diversity of businesses, services and facilities. Business diversity that corresponds to social diversity is most likely localized, or at least small in scale. While it may make sense to talk about economic diversity that fosters innovation and thrives on creativity in the sense of a 'creative economy', such diversity needs to be meaningful to individuals rather than corporations and city politicians. The diversity required is likely not in the form of 'town centers' or other attempts to commodify a nostalgic notion of the American small town. Mixed businesses for a diverse neighborhood need to be something more fundamental. In addition, the composition of the mix matters. Uses should compliment each other and be active at different times of the day, creating what Jane Jacobs called 'complex pools of use'.

The stylistic implications of this kind of mix can be complicated. On the one hand, architectural variety is valued as a way of supporting diversity. Many urban planners and designers reject the uniformity of modernist urbanism, and promote instead an approach that respects variety. On the other hand, uniformity of design can be culturally rich. Sometimes urbanism has a regularity and uniformity to it that provides a supportive context for an underlying social diversity.

Diversity and contrast go together. The key to making the contrast 'work' is context. Gehry's Bilbao museum has been heralded as a building

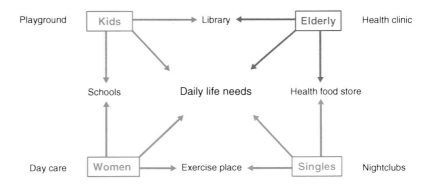

A complex set of services and facilities are required to keep a diverse population sustained.

with extreme contrast that is nevertheless highly successful because of its positioning at the end of a street vista (Duany *et al.*, 2003). It is contextualized by the way the buildings frame it. Similarly, diverse building types reflective of a diverse place may be contextualized in different ways.

Small, independently owned businesses in diverse neighborhoods are essential.

ASSESSMENT

There should be an initial assessment: what does the neighbourhood have? Does the mix of services and facilities meet the needs of the mix of people located there? What groups are over or under serviced? Obviously, many needs can not be met from within the neighborhood; but what could reasonably be added?

One way to support mix is to find areas with high levels of social mix and make sure these areas are well served by public space and/or commercial and other types of non-residential facilities.

FITTING IN SMALL BUSINESSES

Services should be well-distributed to reach a broad geographic range. This also implies that mix has a relationship to size. Many urbanists have stressed that the most important requirement for good urban places is the control of size – keeping things small. Smallness is especially important in diverse areas because it is the basis of multiple ownership, which in turn, encourages diversity of service. Small, independent businesses in diverse areas need to be protected and nurtured, and new small business growth encouraged.

Accommodating the growth of smaller businesses can be accomplished in multiple ways. Although the diverse neighborhoods studied here are built out, there are lots of places to fit in small businesses – for example, adjacent to alleys and near existing commercial buildings. Areas that are already highly mixed in terms of land use should be targeted.

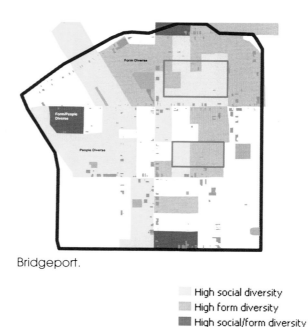

Bridgeport.

High social diversity
High form diversity
High social/form diversity

Irving Park.

West Ridge.

Portage Park.

The red boxes on these maps are areas that are high in diversity but low in terms of facilities. Designers should look at these in detail.

The architect Michael Pyatok argues that designers should help accommodate small, family-run businesses. For lower-income and 'blue-collar' families, homes should be able to function as workplaces – for making clothing, operating a repair shop, or engaging in other types of non-noxious, light industries.

Alleys can be used to support these kinds of family-operated small businesses. An important point is that most of these diverse neighborhoods already have an intricate system of alleys. These need to be legitimized as places for small business – perhaps limited to places near commercial districts and/or places that already have a high degree of land use mix.

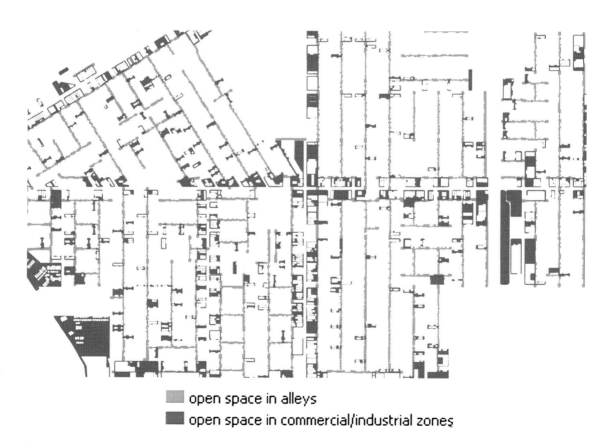

■ open space in alleys
■ open space in commercial/industrial zones

Diverse neighborhoods often have a dense network of alleys and vacant land adjacent to commercial buildings – for example the green and purple spaces shown here in Bridgeport. These areas may be able to accommodate additional small business growth. Unoccupied commercial and industrial land in the spaces between and behind buildings should be tapped for their potential to accommodate small entrepreneurs. All of the areas shown here are within a very short walk to a bus stop, which lessens the pressure for additional parking.

System of alleys in Bridgeport.

The network of alleys in Bridgeport could make it possible to integrate small family-run businesses in the residential fabric, while maintaining a socially functional street. Form-based codes can be used to ensure compatibility and minimize friction.

BIG BOXES

While small neighborhood-based establishments serve and employ a diverse population, large retailers are a mixed blessing. They provide the neighborhood with needed retail and employment, but they disrupt connection, often degrade the collective capacity of the street, generate traffic, inhibit access to all but car owners, and can displace the independently owned retailers essential for diversity.

Big boxes and chains can be accommodated in a way that does not work against the fabric of a diverse neighborhood, which tends to be finely grained. One strategy is to 'wrap' them with smaller units. Or, similarly, put small retail spaces in front of big boxes. Providing spaces for smaller, local retailers is good for diversity, and good for activating and enlivening the street. As Jacobs argued, 'Streets need controls ... but the controls needed are not controls on kinds of uses. The controls needed are controls on the scale of the street frontage permitted to a use' (Jacobs, 1961).

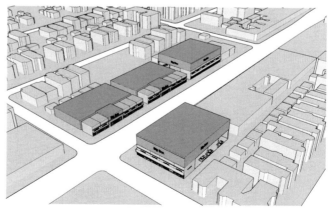

There are ways to fit in bigger retailers without disrupting the smaller scale of businesses essential for a diverse neighborhood.

Encouraging live/work units, artists' lofts, and light manufacturing in the underutilized land adjacent to major transportation corridors may be good for diversity. It helps small businesses and entrepreneurs gain a foothold (or retain their presence) in a diverse community. Such places can function as low-rent business incubators.

The two views above show new infill units (in orange) in a vacant section near the highway in Bridgeport.

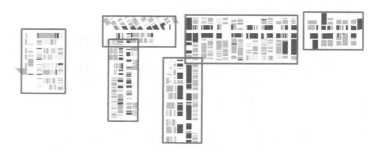

Target areas that have an especially high degree of land use mix, such as the areas in the red boxes, above.

In these targeted areas, there is room to accommodate new small businesses – the darkened green and purple areas are open land.

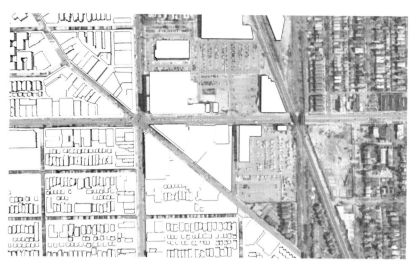

Existing.

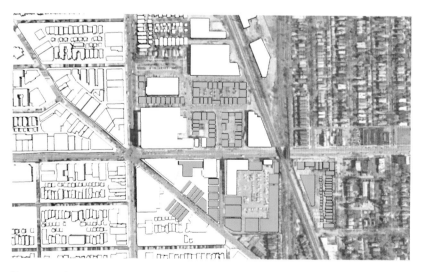

Proposed.

An underutilized area near the highway in Portage Park could be used for small business development.

Around a major rail corridor in Bridgeport, there are parking areas that could be used for artists spaces, modular manufacturing buildings, and prefab units of various kinds.

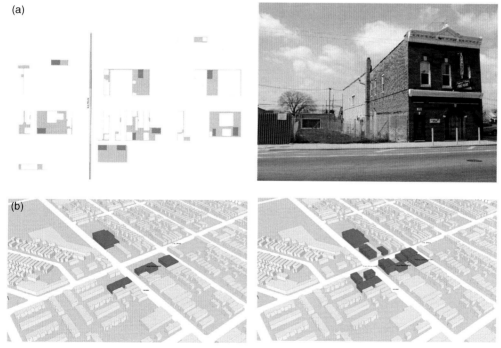

(a)

(b)

Existing.   Proposed.

Along commercial corridors, which are also well served by public transit, there are empty lots to fill in with small buildings – buildings to accommodate the varied services and facilities diverse neighborhoods require. These could be tucked in and made unobtrusive, as at (a), or positioned to frame the street, as at (b). There are multiple benefits in both approaches, including added surveillance of underutilized vacant spaces, and the ability to frame space that is otherwise given away to car traffic.

Around industrial land and transit corridors – which diverse neighborhoods often have plenty of – there is space to fit in small businesses: resilient, small, flexible units to accommodate small entrepreneurs. In these areas, codes should be relaxed and incentives given to develop the kinds of small businesses that are essential to maintaining a diverse neighborhood. It should be easy for small starter businesses to locate near industrial sites and railroad tracks – leftover spaces that often go vacant and underutilized. Because of the fact that small businesses are essential for sustaining diverse neighborhoods, public subsidy in the form of tax breaks and low-interest loans should be instituted.

The charrette team developed a number of designs for easy to build, easy to expand, relatively low-cost commercial structures in a vernacular neighborhood character.

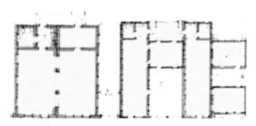

The charrette team designed these floor plans for the commercial structures.

Small modular units can be used to set up shops easily. These small retail and artist spaces are designed to be affordable, small, and neighborhood-based. Above left: retail spaces in New Town St. Charles, MO designed by Duany, Plater-Zyberk & Co. (photo by Sandy Sorlien); right above and below: easy-to-build commercial structures from the Gentilly neighborhood plan (a section of New Orleans, by Duany, Plater-Zyberk & Co., p. 84); below left: a cluster of pre-fab artist spaces from the Bay St. Louis rebuilding plan (*Mississippi Renewal Forum Report*, p. 11, courtesy of Michael Mehaffy).

8 Connection

The importance of maximizing connectivity in urban space is a common theme in urban design (Alexander, 1965; Hillier and Hanson, 1984). Salingaros (1998), for example, has written that the 'living city depends on an enormous number of different paths', and that 'dead cities need to become more connected to regain urban vitality'. Common design strategies for promoting connectivity include gridded street networks, establishing the '100 percent corner' where multiple activities can coalesce, or providing well-located schools, parks and neighborhood stores as shared spaces that foster social connectedness. Connecting all types of spaces is important – public and private, residential and non-residential, storefront and sidewalk.

To achieve connectivity, planners may focus on the alternative routes and access points that can be created by increasing street connections. They may draw attention to the size and shape of blocks, which determine both the public space network and the corresponding patterns of movement. It is generally agreed that large-scale blocks, cul-de-sacs and dendritic (tree-like) street systems are less likely to provide good connectivity.

Enhancing connectivity can also be as simple as delineating safe places to cross existing streets, calming traffic down on busy streets, or instituting better pedestrian pathways. Most of these strategies are based on the view that the built environment can have an effect on constraining or promoting passive contact, and that social interaction may ultimately be tied to the amount of passive contact that takes place (Fischer, 1982; Gehl, 1987). It is also known that human interaction at the neighborhood scale is a pedestrian phenomenon (Michaelson, 1977), and that 'networks of neighborly relations' are related to networks of pedestrian streets and the internal neighborhood access those street networks engender (Grannis, 2003).

Public places, because of their connective ability, play an especially important role in sustaining social and economic diversity. No one

Treatment of key intersections is often weak and has little value for enhancing neighborhood identity, such as this intersection along a main commercial street in Bridgeport.

Two important intersections in Bridgeport, shown at left in the red boxes (and above and below), have little value as identity space – essential for connection in a diverse area.

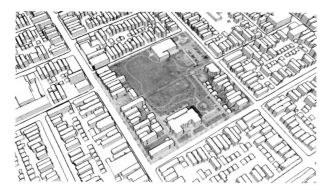

This strategically located area at the heart of Bridgeport is occupied by a playground and park, but unfortunately, neither function as a point of community connection. The park is closed after school hours, is only accessible to school children, and lacks formal entranceways.

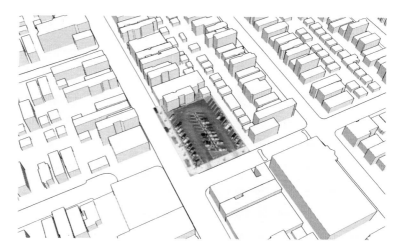

This strategic location and potential connecting point in Bridgeport contains a parking lot, which has little value for creating community identity.

would say that the provision of shared space is all that is needed to make diversity work, but it is certainly a critical element. The provision of public spaces for casual or spontaneous interaction does not create deep social bonds, but instead promotes 'weak' social ties, which are not only necessary and important, but are believed to be especially sensitive to environmental design (Skjaeveland and Garling, 1997).

To some extent, loss of diversity mirrors the rise in privatized realms. The public restaurant is replaced by the private indoor kitchen, the public playground is replaced by the private back yard, and American industry is sustained by ensuring that the masses are housed in places large enough to substitute for a community (Oldenburg, 1999). Given this corporate backing, re-establishing a well-functioning public realm as an essential aspect of sustaining diversity will not be an easy task. The approach will need to be incremental and small scale. Accordingly, the design strategies suggested here are relatively modest.

I focus on four aspects of connectivity that are relevant to an existing diverse neighborhood (as opposed to a new development where laying out street grids and insisting on short blocks may be more relevant). I call these: Identity space, collective space, institutions, and networks.

Two views of what could be called the 'civic deserts' of Portage Park. The map on the left shows areas that are not covered by any connecting space. The areas marked as 'civic deserts' are blocks that do not come close to parks (green), schools, libraries, community centers (blue dots), or commercial facilities (dark purple). The right map shows these deserts in relation to the areas most diverse in terms of form (red) and people (purple). Designers should focus on enhancing areas where civic deserts overlap with high diversity.

To locate neighborhood centers, find places where there is a convergence of activities. The map of Bridgeport on the left identifies commercial parcels, libraries, senior centers, schools and parks. The map on the right narrows the search to include only locations near mixed housing type and the busiest intersections. These locations should be investigated as potential sites for neighborhood centers.

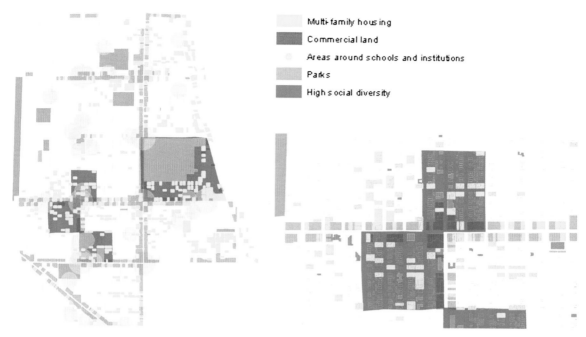

Multi-family housing
Commercial land
Areas around schools and institutions
Parks
High social diversity

Another strategy is to start with the same criteria as above, but narrow the search to focus on areas with particularly high levels of social mix (areas marked in red). Closer in views (above right and below) in West Ridge show land use diversity and vacant land available for development. These locations have the potential and the space to function as centers. It is not uncommon for key locations in diverse areas to include vacant, underutilized land or parking lots.

For each strategy, a primary goal is to find strategic points for making connections and locating something of value there. In addition to strategically targeting improvements, the strategies involve small, incremental changes.

IDENTITY SPACE

The creation of some sort of identity for a diverse place is important, possibly crucial. The images, symbols, and landmarks of a neighborhood serve to hold a diverse population together and provide a rallying point. Identity space provides a way of binding disparate people and places. It is related to what Paul Zucker defined as the 'nuclear square' – a place that pulls heterogeneous elements at the periphery together, into a visually integrated unit (Zucker, 1959).

A more focused core is one way of creating a shared identity. Often, there is no centralized, accessible space that represents the community as a whole. At the geographic heart of the community, there may be only a commercial corner with little attention to design or its effect on the public realm. Central corners on major streets should not consist of empty lots.

Diversity is regularly thought of as something that emanates from a center (for example, Girling and Kellett, 2005). An initial assessment should be made to determine what the geographic 'heart' of the neighborhood is. Where are the centers or subcenters? Is it possible to identify potential neighborhood centers where more intensification could reasonably occur, including especially multi-family residences? Perhaps the identity space is not located in one centralized or large area. It could be a series of smaller ones.

It should be determined whether there are spaces where a diverse population crosses paths, and at those points, whether there is a functioning public realm. Look for the convergence of 'layers' of activity: pedestrian walking sheds, major intersections, housing type diversity. At those key intersections, it may be especially detrimental to have vacant land or parking lots. Instead, key intersections at major streets should serve as neighborhood centers and identity space.

Along a main street, a series of weak intersections is especially taxing for enhancing community identity. But parking lots, which have both weakness and potential, can become connective public space in a diverse neighborhood. One strategy is to use them simultaneously as parking lots and public squares, as they do in Europe.

Identify potential subcenters. Start by fnding areas where multiple functions, housing type diversity, and key intersections converge (above, top left). Potential subcenters, such as the red parcels shown in three West Ridge locations above, are vacant parcels adjacent to public and quasi-public buildings. These could be developed as pocket parks, civic plazas, local markets, or other public gathering spaces of various kinds.

A parking lot and privately owned business at the geographic heart of Irving Park, near the El station.

In the heart of Portage Park, across from the park that bears its name, a parking lot for a convenience store.

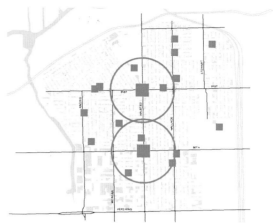

A parking lot and a gas station at the two most central locations of Bridgeport.

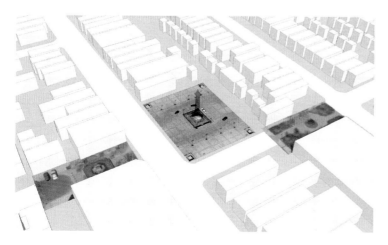

With the addition of better paving materials and an identifying feature like a fountain, this parking lot (left) at a main intersection in Bridgeport could be adapted to function as a public space while still providing parking. The two adjoining public areas are too small to function as a neighborhood center. The parking area could function as a neighborhood center if it were connected to the other public spaces and redesigned to function as public space.

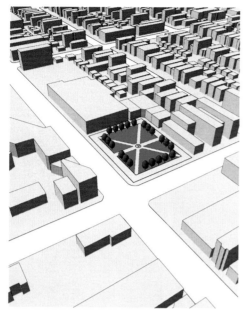

At the heart of a diverse community, there should be something that connects people, by providing 'identity' space. Above is the geographic heart of Bridgeport, below the geographic heart of West Ridge. The design on the right above and below are insertions from a civic plaza design found in Hegemann and Peet's (1922) *Civic Art* (p. 148).

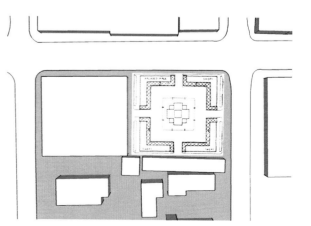

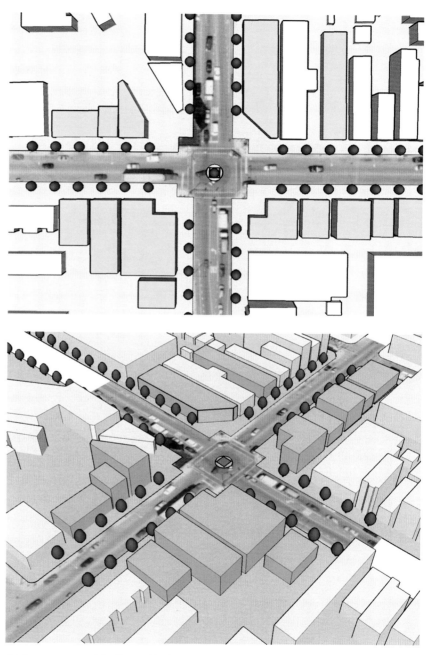

At the geographic 'heart' of each community, there is opportunity to enhance the public realm and create a more meaningful center.

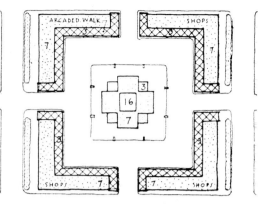

Strategic corners should have a dignified design. Along the main commercial corridor of Irving Park, there currently sit car dealerships and a McDonalds. Giving central locations some dignity through design may have great connective value. The arcaded design above right is inspired by Hegemann and Peets in their 1922 *Treatise on Civic Art* (p. 148).

The main entrance to Portage Park (built 1913) is dignified and stately.

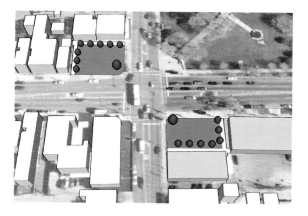

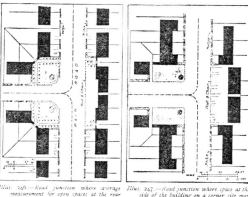

Illus. 246.—Road junction where average measurement for open spaces at the rear is allowed.

Illus. 247.—Road junction where space at the side of the building on a corner site may be substituted for space at the rear.

The strategic corner across from Portage Park is currently occupied by convenience stores and car dealerships (above right). The proposal above left is inspired by an intersection design of Raymond Unwin, shown above, found in his 1909 *Town Planning in Practice*, p. 334.

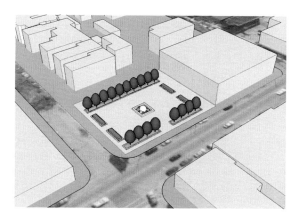

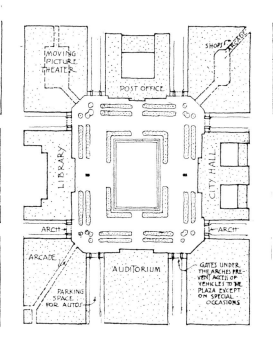

A strategic corner in West Ridge, currently the site of a gas station, can be of very high strategic value for a diverse place. This design is inspired by the civic square designs of Hegemann and Peets, such as the one on the right (*Civic Art*, p. 148).

COLLECTIVE SPACE

Collective space is less about forming an identity and more about finding opportunities for interaction. Collective spaces promote exchange. This is important in any neighborhood, but it is especially important in a diverse neighborhood as a way of counteracting the distrust or fear residents might be harboring about people unlike themselves. Sequestering, enclaving, barricading – isolationist strategies are not sustainable coping mechanisms in the long term. Collective space promotes interaction, providing a better chance for informal, voluntary control.

Points of connection should be woven into everyday movement and activity space – like parks, public schools, and libraries. Ideally, if these spaces are woven throughout in a successful way, all residents can be within walking distance of a connective space of some kind. It is also important to have collective spaces that don't have a specific agenda or goal, but can be used by multiple people for multiple purposes. On the other hand, collective space specifically targeting children – schools and playgrounds – is essential for maintaining socially diverse neighborhoods, since children have been found to be an essential ingredient in forming social connections (Gleeson and Sipe, 2006).

Collective space can be fitted into small spaces. Small grain is important since frequent and well-dispersed public spaces build connection within the context of everyday activities and movement. (This is very different from newer suburban areas, where civic spaces tend to be large and only accessible by car.) Empirical studies have demonstrated that users will frequent public space most often if they can walk to it, and, if it is within 3–5 minutes walking distance from their residence or workplace (Kaplan and Kaplan, 1989).

Collective space should be located near places with a high level of diversity. Ideally, they should be located at points of intersection between well-traveled streets, or near institutional buildings. If they can be located on vacant land, or vacant commercial buildings, there may be greater chance of implementation.

There may be opportunities to add all sorts of public, quasi-public or private non-residential uses that function as collective space. The strategy would be to encourage mixed use in these selected locations. These new spaces would not necessarily be intensive enough to support neighborhood-level retail, nor would they supplant the commercial function of the main street corridor. But they would provide non-residential focal points to support the diverse surrounding area.

Special attention should be paid to the urban form surrounding collective space. Unfortunately, many public facilities are surrounded by 'dead space' – parking lots, for example. This breaks their connection with the surrounding social fabric, and impedes their ability to function as connective spaces for a diverse population.

Streets as connecting space

Using streets as connectors requires reading a streetscape as a habitable space rather than a conduit for moving cars. Streets have an

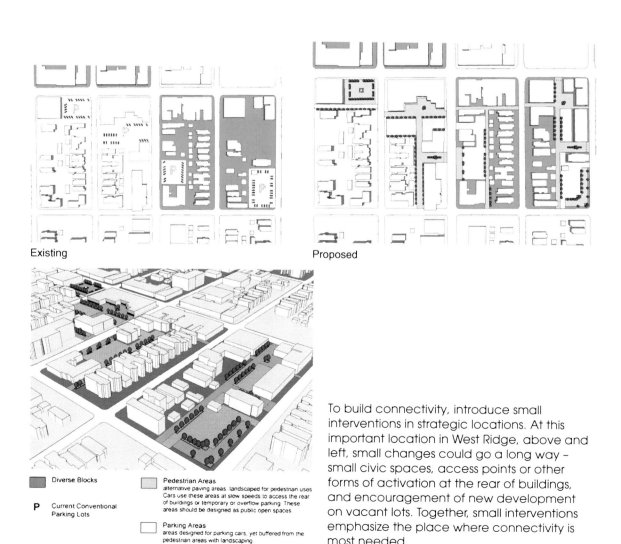

Existing   Proposed

| | Diverse Blocks |
| **P** | Current Conventional Parking Lots |

Pedestrian Areas
alternative paving areas, landscaped for pedestrian uses. Cars use these areas at slow speeds to access the rear of buildings or temporary or overflow parking. These areas should be designed as public open spaces

Parking Areas
areas designed for parking cars, yet buffered from the pedestrian areas with landscaping

To build connectivity, introduce small interventions in strategic locations. At this important location in West Ridge, above and left, small changes could go a long way – small civic spaces, access points or other forms of activation at the rear of buildings, and encouragement of new development on vacant lots. Together, small interventions emphasize the place where connectivity is most needed.

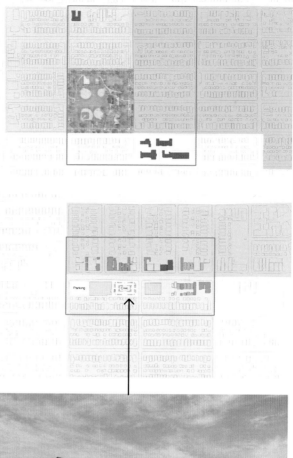

Building connectivity in a diverse neighborhood involves finding strategic locations for interventions. In the two sections of Portage Park above, the areas shaded red have high population density, while the areas shaded in green have high social diversity. In the transition areas between these two critical demographics, connection is important. The section of Portage Park on the left is a positive example – a small neighborhood park sits at the juncture of high diversity and high density. Civic buildings are shown in blue. On the right, connectivity could be strengthened by inserting more thoughtful civic spaces in at least one of the two parking areas.

obvious effect on separation and the disruption of neighborhoods. A recurring phenomenon in diverse places is overly busy thoroughfares – streets with six lanes of traffic buzzing through the center of a diverse community. Such roads have value as promoting external connectivity, but there has been a cost – disruption of pedestrian quality and thus connectivity.

The importance of streets as elements of public space has been a common theme in urban design for quite some time (see especially Southworth and Ben-Joseph, 2003). If streets can be conceptualized as a form of public space, they can act as linkages between otherwise separated places. In well-traveled areas, ample sidewalk width and street trees could be used to buffer pedestrians from cars and enhance the ability of these areas to function as collective space.

To better define the public area, buildings could be encouraged to form a disciplined frontage and thereby give a better sense of the public realm. Planners can encourage this by installing a build-to line on strategically selected blocks, which is simply a way of coding uniform setbacks. A 'build-to' line can help create a better sense of enclosure on the street, thus helping the street maintain its function as an important, connective space. This technique may also help to encourage land use diversity by smoothing out its potentially chaotic effects (and lessening opposition to it).

INSTITUTIONS

Institutions played a large role in sustaining the social diversity of 19th and early 20th century neighborhoods. There was an infrastructure that supported social diversity: local churches, local schools, local retail like corner markets all contributed to a dense network of support for a diverse population. Alexander von Hoffman's (1994) study of urban neighborhoods between 1850 and 1920 showed how economic interdependencies fostered attachment to others and to the neighborhood. Grocery stores and hardward stores served as community centers. We can see remnants of this past hey-day of social infrastructure in diverse neighborhoods now.

The institutional resources of a community – what kind and how they are integrated – is an essential aspect of designing for diversity. There

Two areas in Bridgeport, outlined in red at right, are simultaneously high in population density and high in social diversity. Some design strategies could be used to strengthen connectivity in these areas.

WOONERF

The image at right shows the housing type mix for the uppermost area above. Since it is predominantly a residential area, connecting space in the form of a woonerf (an example was shown in Chapter Seven), would be particularly appropriate. Note the small amount of civic space (building footprints, in black) that currently exists in the area.

one-story homes
two-story homes
small multi-family like duplexes
larger multi-family and apartments
commercial buildings

■ Civic buildings

Parks often function like institutions because community classes and after school programs are run through park district buildings. Yet the geographic distribution of parks is not always conducive to fostering connection. The red areas in West Ridge, above, contain the most social diversity, but some areas are not well served, or are served by large, passive recreational spaces that are less likely to function as institutional connectors.

shouldn't be a burden involved in getting to these institutions. They should be pedestrian-friendly, well connected, reachable by public transit, and free of excessive congestion – accessible to the walker, biker, transit-rider and the driver.

Evidence of an infrastructure that once supported social diversity: churches, schools, theaters and corner markets all contributed to a dense network of support for a more heterogeneous population. Because of their small scale, they could be geographically dispersed. The theater shown above left is located along Portage Park's main street and is currently under renovation.

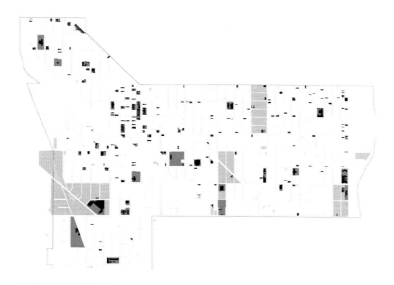

The purple areas on the maps above and left are the most socially diverse parts of Irving Park. Non-profit institutions are shown in the red spaces. The spatial pattern of institutions seems to bear little relation to the most diverse areas – those likely to be in most need of institutional infrastructure.

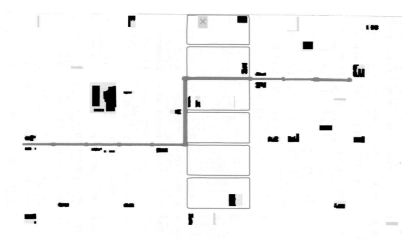

There are several strategies for increasing connectivity via institutions. First, target scarce vacant lots or buildings in the most socially diverse blocks (outlined in red, at left). Vacant parcels are shown in yellow; the red 'x' is a larger parcel suitable for institutional development.

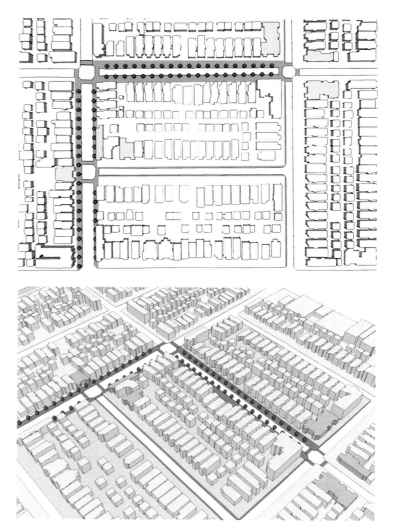

Another strategy would be to build connectivity between institutions – institutions (in blue) along the red line (above) could be networked by targeting streetscape improvements along the route that connects them. This has been used as a way of connecting light rail stations: using a paving pattern that goes from station to station through the neighborhood as a way of integrating the entire system. Similarly, Edmund Bacon wrote in *Design of Cities* (1976) that it was the movement systems between points that constituted the organizing framework of Baroque Rome.

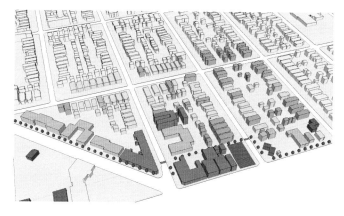

Crosswalks and trees could connect the various institutional uses (in blue) in this diverse section of Irving Park.

Design strategies to strengthen this network include adding improvements to the open spaces around the institutional buildings so that they are designed for the purpose of connection (see example, left). If the institutions are within the same block, it may also be possible to better connect institutions to form a mutually-reinforcing network. Adding modest street and sidewalk improvements between them would help accomplish this.

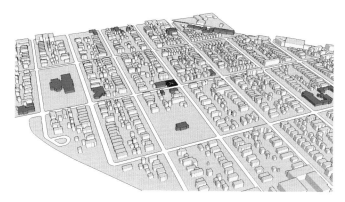

This section of Irving Park has a well-dispersed network of institutions (red buildings).

This area around a school in Irving Park is a well positioned neighborhood connector.

The institutions shown in light blue are embedded in an area (in red) that has a very high level of 'form diversity' – mixtures of housing types, tenures, and sizes. The school and surrounding space is well positioned.

Institutions that do exist should be encouraged to design quasi-public spaces that encourage connection. This church in Irving Park is centrally located in a highly diverse area. Ideally, the small park should be designed to function as a community connector – by adding a few benches, and giving people the choice of either passing through or lingering. The inserted park design, below, was inspired by a drawing in *Designing Small Parks* by Ann Forsyth and Laura Musacchio (2005).

Existing

With park redesign

NETWORKS

Residents should be able to move in multiple ways throughout the neighborhood. This is not only more convenient for pedestrians, but it increases opportunities for social transaction, likely to be a good thing

Unsociable spaces don't just impact the space in front of them. Together, they can disrupt the whole neighborhood network. The above map of Bridgeport shows vacant land in black and the street segments the vacant parcels are connected to in red. A large portion of Bridgeport is therefore connected to 'unsociable space', weakening the overall network.

for diverse neighborhoods in the long-term. Dead-ends and cul-de-sacs disrupt the street network. If there must be a dead-end, it should end with public space and a pedestrian pathway. (This was one of the design innovations of Radburn, New Jersey). Also potentially detrimental are dead-end streets, sidewalks that stop abruptly, parking lot frontage and other blockages to the interconnecting paths of a neighborhood.

Empty spaces along the street network of a neighborhood can also have a negative effect. Vacant lots don't just impact the space in front of them; in aggregate, they can disrupt the whole neighborhood network. Streets are social connectors, seams that bind. In many places, these seams don't bind well because of the empty spaces on either side. Putting back the 'missing teeth' is one way to repair the seams.

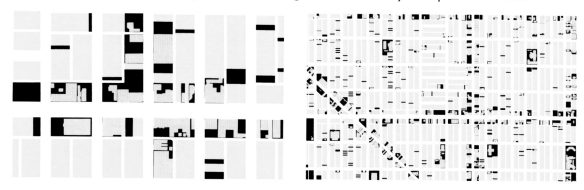

These maps are sections of Irving Park. As with the other diverse neighborhoods, Irving Park is covered by a well connected street grid, which is ideal for connectivity. However, there are disruptions in the network – wide, busy streets, parking lots, dead ends, and vacancies.

The black areas and blue buildings in these maps represent vacant space around non-residential buildings. In these locations, designers should put in small interventions to help improve connectivity – benches, painted or textured sidewalks, street trees (one simple example is shown below). The idea is to keep the sidewalk alive in the most strategic places – where there is diversity all around.

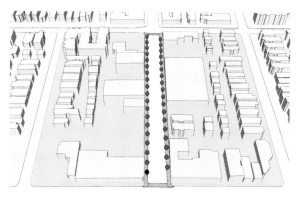

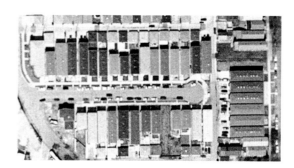

Dead-end streets like cul-de-sacs, above, disrupt connection because there is no way to pass through. If there must be a dead-end, it should end in public space, as at left. Both examples are from Bridgeport.

These maps of West Ridge and Bridgeport show the blocks with the highest diversity (purple) and the highest density (red), relative to the buildings along their main streets (green).

The maps show that, traversing these highly diverse and dense blocks, there are empty spaces that disrupt the ability of the street to function as a defined space and a social connector.

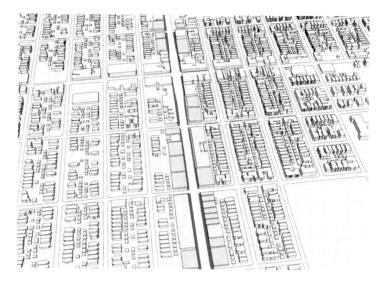

A 'build-to line' can help create a better sense of enclosure on the street. It helps the street maintain its function as an important part of the connective public realm. This street is in West Ridge. The yellow buildings are on sites that are currently vacant or used for parking.

9 SECURITY

The issue of security is a third dimension in diverse neighborhoods that designers can respond to. It goes beyond Jacobs' notion of 'eyes on the street' and CPTED strategies (crime prevention through environmental design, Jeffrey, 1971), important as those are. If people and functions are to be mixed and connections between them are to be enhanced, security needs to be addressed in multiple ways. Above all, designers will have to be sensitive to the need for people to feel secure about the mixing and connecting required of them in a diverse place. The solution is not seclusion and withdrawal, because that only breeds additional fear. There must be a balance between social mix and connection and the essential requirement that people feel safe and secure in diverse environments.

There is no denying that diversity possesses the inherent paradox of providing both a source of innovation and creativity, and, a source of tension and conflict. The two effects are interrelated. The urban designer is left with two tasks: fostering and sustaining social mix, and, counter-acting the negative effects of that mix – i.e., the conflict and fear that impedes rather than enhances neighborhood social and economic functioning. Design is crucial especially since it may not always be possible to rely on long-term resident commitment as a foundation for security.

Diversity in urban places is thwarted by the constant fear that 'low status' people and uses will hurt 'high status' people and uses, and vice versa. This happens on multiple levels: the highest status fearing the lowest, but also intermediate levels fearing levels immediately 'below'. The latter condition may be the most endemic. As Kefalas (2003) found, 'working-class homeowners' socioeconomic vulnerability heightens awareness about the risks created by living in close proximity to lower-class populations' (p. 100). Perhaps attention to design can lessen these concerns.

New developments in diverse neighborhoods are sometimes monolithic and shielded from the street. New housing development can seem unsociable, especially if the buildings have no openings onto the street. In the long run, this reduces security because it severs connection to, and responsibility for, the public realm.

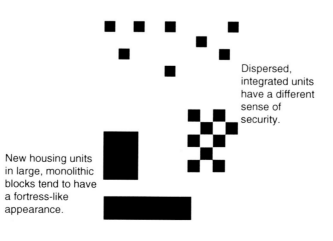

Dispersed, integrated units have a different sense of security.

New housing units in large, monolithic blocks tend to have a fortress-like appearance.

Sometimes these fears are manifested in attempts at separation and seclusion. There may be physical indicators that people are barricading themselves in – fences, gates, setbacks, fortresses. This practice undermines connectivity. The solution is to build in security in ways that don't do harm to mix and connection.

I look at four ways to address security through design in diverse places: housing integration, surveillance, activity and edges.

HOUSING INTEGRATION

There seems to be a growing prevalence of housing enclaves in diverse neighborhoods. There are high-rise condos where owners drive into the back of their building and never come out of the front door, 'townhomes' with gated entranceways, and new single-family housing in separated, monolithic pods. Housing enclaves – developments turned inward, with restricted access – are most likely born out of a desire to detach from the surrounding neighborhood in the hopes of increasing security. Yet if every development were to act defensively, the net effect would be decreased security because no one would be paying attention to the surrounding streets and public spaces. Enclaves that are sequestered, fenced, or have anti-social frontages of various kinds have the effect of disrupting connectivity and ultimately, decreasing security.

New housing developments should be integrative, not walled off and abruptly insular. While new housing development obviously comes in grouped form, there should be a portal of some kind, a meaningful opening or connection to the surrounding neighborhood.

One example is housing around looped streets (called a 'close'), as shown in Chapter Seven. Such an arrangement allows communal space shared by adjacent dwellings, but there is also sufficient integration with surrounding single-family homes. Techniques like the close integrate a variety of housing types in proximity. If housing must be set off, it should make use of entrances that connect, not sequester.

Residential enclaves weaken connectivity and may lessen
security. The development above is walled off from the street.
In contrast, the homes below connect directly to the street and
therefore – potentially – take more responsibility for it.

New housing developments are fitted in as separate enclaves, like these in Portage Park, above and Bridgeport, below.

Each of the figure grounds below show a section of the neighborhood that could benefit from infill housing that is integrated with its surroundings, not walled off to produce a false sense of security. As a practical matter, this involves suggesting to developers new, more creative approaches to infill housing, such as employing the historical building types shown here.

Existing 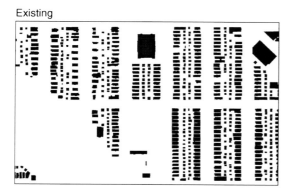 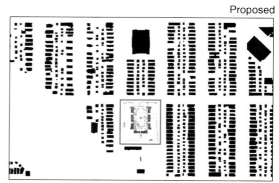 Proposed

Irving Park offers the perfect opportunity for infill housing that is connected to its surroundings, not barricaded and closed off. Inserted housing layout is from Raymond Unwin's (1909) *Town Planning in Practice* (p. 353).

Existing 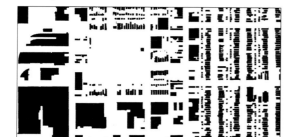 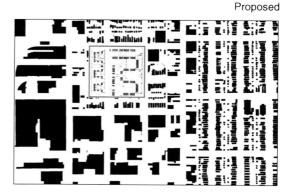 Proposed

In Portage Park, there is a need to find a building fabric that can transition well and tie together all the different typologies found. Inserted housing layouts are from Raymond Unwin's (1909) *Town Planning in Practice* (pp. 330, 353 and 350).

Existing

Proposed

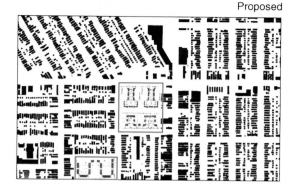

Bridgeport enclaves in the form of superblocks can be replaced by housing that integrates better with its surroundings. Inserted housing layouts in two locations are from Raymond Unwin's (1909) *Town Planning in Practice* (pp. 330 and 353).

Existing

Proposed

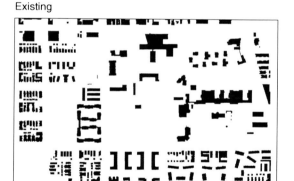
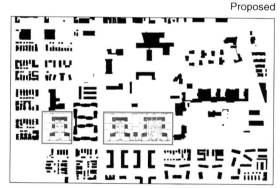

There are many different building types and forms in the West Ridge neighborhood shown here. Separating each type does not necessarily increase security. Innovative building forms can provide a sense of security but still connect to the surrounding area. Inserted housing layouts are from Raymond Unwin's (1909) *Town Planning in Practice* (p. 353), and Hegemann and Peet's (1922) *Civic Art* (p. 266).

SURVEILLANCE

It is possible to enhance the surveillance, control and responsibility for public spaces. This was a key dimension of Jane Jacob's (1961) argument for successful urbanism – that there should be natural surveillance built in. It should be relatively effortless for people to keep an eye on things, as part of their everyday routines. Design plays an essential role. Urban structure can make it easy for people to focus their 'eyes on the street' – paying attention to, and taking responsibility for, the security of the neighborhood.

To promote natural surveillance, buildings should front public space. People should be able to look out of their windows directly on to the public realm in front of them.

Natural surveillance of public places like parks is essential, especially in a socially diverse area. Parks should not be fronted by garages, parking lots, or the sides rather than fronts of buildings. Instead, spaces around public places like parks should front the park directly and engage with it. This dignifies park space, acknowledges its social value, and increases security.

Because of the way the surrounding residential fabric is designed, this park in West Ridge is detached from its surroundings. Infill development should be designed to engage with the park more directly.

Proposed infill development shown in red

This park in Bridgeport, in the heart of one of the most socially diverse areas, is currently fronted by garages, parking lots, and the sides rather than fronts of buildings. To allow natural surveillance and promote the park as a connective space, new buildings should face the park directly.

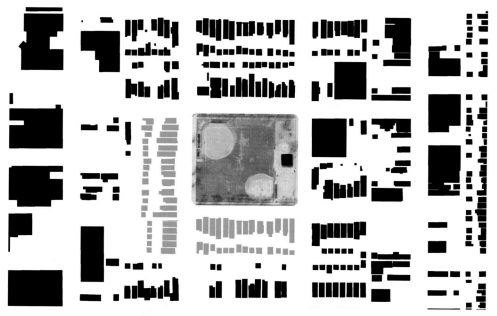

Proposed infill development shown in red

Planners should provide incentives to increase park-facing infill development surrounding this park in Bridgeport, currently adjacent to empty space on two sides. This will activate and strengthen the park as an important part of the public realm.

This park in West Ridge is well integrated with its surroundings – apartments and single-family homes face directly onto the park, offering a natural form of surveillance that increases security.

ACTIVITY

Another aspect of using design to increase security is to activate 'dead' space – empty, unclaimed or underutilized land for which no one seems to be taking responsibility, and for which there is little security for the passer-by. Empty space is not just harmless open space, or someone's sideyard, or an industrial zone. It is space that fronts the public realm, but has no connection to it. Parking lots are perhaps the largest category.

In Chapter Eight these dead zones were discussed in the context of disrupting connectivity and impacting the broader network of a neighborhood. But relatedly, dead zones decrease security because they decrease levels of activity. This is especially true in commercial corridors. In the main arteries of the diverse community – the important commercial corridors running through the neighborhood – there are gaps that de-activate. Short of having 24-hour police patrols, the way to build up security in these areas is to make sure they contain active uses – not parking lots or vacant land.

Keeping neighborhoods 'alive' with 24 hour police stations, like this one in the center of Bridgeport, may not be the ideal way to provide a sense of security in the heart of a neighborhood.

Strong commercial corridors are essential, and residents and visitors have to feel secure in them. In addition to parks, the commercial arteries of these communities constitute the primary public realm. To strengthen security, the spaces of commercial corridors should be lined with active uses rather than 'dead space' like parking lots.

Too much space in this main commercial section of Irving Park is given over to cars and the storage of cars. In these sections alone, there are 440 spaces, as shown. Each block contains 25 to 50 parking spaces.

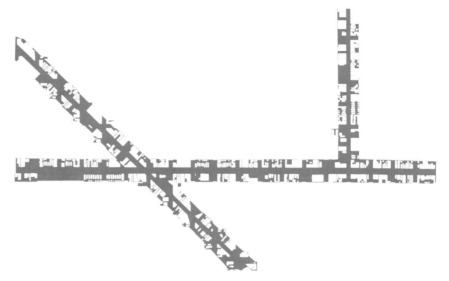

The main commercial corridor of Irving Park has too much 'dead space' (shown here in red) – inactive space between buildings or given over to cars.

Photo courtesy of the US Environment Protection Agency. Reproduced with permission.

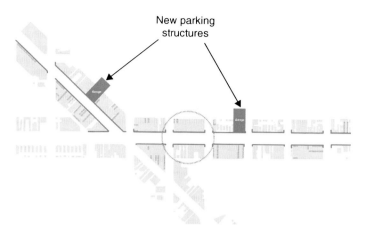

New parking
structures

To activate the commercial core and therefore increase security in this section of Irving Park, three design strategies could be used. First, replace surface parking lots with parking garages. As shown here, the spaces given over to surface parking could be accommodated in two parking garages, and still be within a short walk of the entire commercial area shown. Specifically proposed here are two 4-story garages with approximately 250 cars each, which would not only accommodate all existing surface parking but, if located as shown, would be within two blocks of all existing parking. Garages in the area do not need to detach from the public realm, but can be lined with commercial space, as shown in the photo of Hollywood Video (opposite). Second, traffic calming measures could be instituted in strategic locations (red circle area, detailed below) to reconstitute the area as a public room rather than a traffic artery – also important for building a sense of security. Third, a build-to line could be put in place to give a similar effect – creating a sense of secure enclosure in the commercial area.

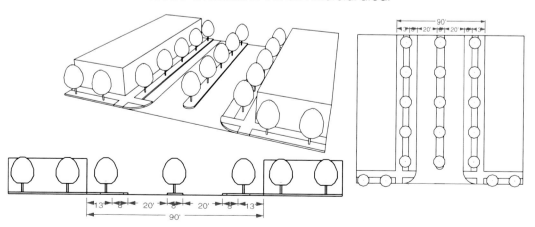

Dimensions of an avenue with pedestrian-oriented design, important for traffic calming. Drawn using dimensions from the SmartCode.

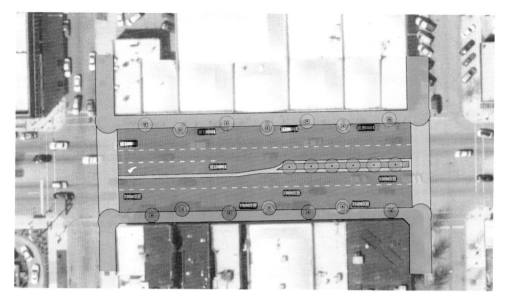

Detail of red circle (on previous page) – a strategic section of the commercial core of Irving Park that could function as a more valued public space if traffic calming measures were installed. The effect would be a greater sense that this area has public value, thus providing a better sense of security.

EDGES

Diverse places tend to have strong edges, composed mostly of transportation and industrial corridors. There are also water bodies, brownfields, and large impermeable districts of various kinds. The question is whether barriers or filters are needed to soften the impact of an undesirable edge. Edges are supposed to bound and give shape and identity (or legibility) to a community, and they often do this. But where the edges are of major proportion and impact, the important consideration is how to buffer them. How can the community be protected from a noisy highway, a barren industrial landscape, or a metallic railyard?

Strong edge conditions – a major transportation route like a highway or an industrial corridor – often play an important role in providing external connectivity, but there is a cost. In and of themselves, industrial corridors near residential areas may not be much of a problem. But highways and rail lines can be problematic for residences, schools, and other non-resilient uses.

Residences can be protected from harsh edge conditions by establishing a buffering greenway, or, by adding resilient building types – like offices or light industrial buildings. Entrepreneurial start up businesses can be accommodated in edge areas as a way of buffering harsh edge conditions.

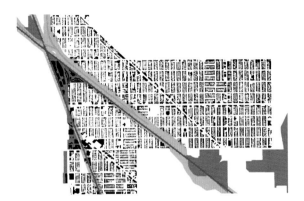

Irving Park edge areas

Bridgeport edge areas

Diverse neighborhoods in Chicago tend to have strong edges running through them. Some mitigation of their adverse affects is important for supporting diversity.

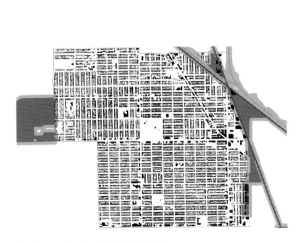

Portage Park edge areas

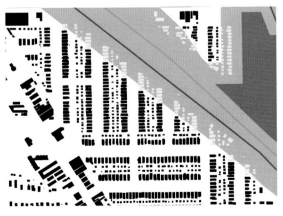

Detail of Portage Park edge area showing the number of residences within very close proximity.

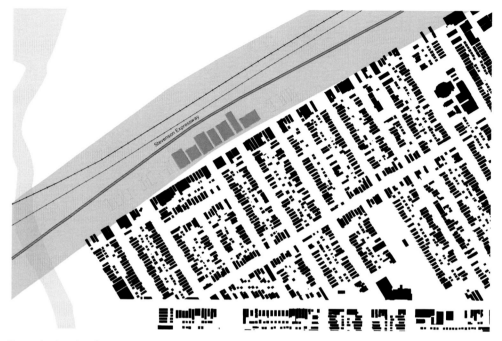

Two strategies for dealing with strong edges include buffering and integration – softening and separating strong edges with a greenway (below), or locating resilient uses like light industrial and office buildings within the edge areas (above).

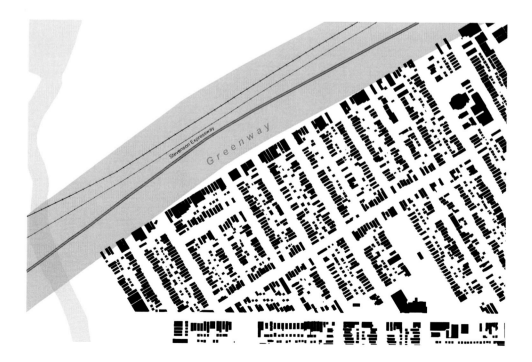

10 CONCLUSION

POLICY AND PROCESS

The viability of stable, diverse neighborhoods in 21st century America sometimes seems perilous. Recently, two of America's most prominent social scientists put the idea in jeopardy. Survey work by Robert Putnam, author of *Bowling Alone. The Collapse and Revival of American Community* (2000), found a surprising level of multicultural intolerance in his longitudinal survey of 26,000 people in 40 communities. He concluded that although diversity is essential for productivity and social well-being, people in diverse places – even non-ethnic Whites – often become more socially isolated. Conservative commentators picked up on this to bolster arguments that multiculturalism is dead (Sailer, 2007). Meanwhile, the Harvard sociologist William Julius Wilson published the results of an ethnographic study called *There Goes the Neighborhood: Racial, Ethnic and Class Tensions in Four Chicago Neighborhoods and Their Meaning for America* (Wilson and Taub, 2006). In it were exposed the worst sorts of racial prejudices. It is an alarming and bleak appraisal.

The Uptown neighborhood in Chicago commissioned this mural to publicize its diversity. This kind of effort could be woven into, and strengthened by, neighborhood planning.

What these studies show is that neighborhood-level diversity and the tolerance and inclusiveness it requires is not, and never was, an easy row to hoe. No matter how enduring the power of a shared American dream, no matter how compelling our historical sense of the moral rightness of American pluralism, we are always going to have to work hard to make diversity in local contexts endure. It is not going to 'just happen'.

I have tried to approach this challenge by looking at the role of the physical environment. I was motivated by the view that, to some extent, the notion of a diverse and tolerant America has been confounded by a physical context that fails to accommodate it. And, I have argued, there are very real ways in which design could be leveraged to provide a much more supportive framework. Attention to design seems to be particularly relevant in areas struggling to hold on to their diversity. As recent ethnographies have revealed, neighborhood newcomers, especially if they are minorities, are likely to be blamed for any deterioration in the physical quality of a place (Wilson and Taub, 2006). This makes attention to design much more significant, not as a strategy for keeping places homogenous, but as a strategy for sustaining diversity.

However, this book did not address the policy, programmatic and process requirements for sustaining diversity. There should be no illusions: it would be impossible to retain diversity via design without giving adequate attention to those issues. The simple truth about neighborhoods that possess the kind of physical design that supports diversity – good connectivity, mix of uses, a diversity of housing types and styles, dignified public buildings, what Fishman (2005, p. 363) described as 'precious legacies of long-lost civic idealism' – is that they have a highly desirable kind of character, vitality and location. Supply does not match demand, and many people and businesses do get pushed out.

Of course, the change from diverse to non-diverse is not always toward affluence. Some diverse places suffer from gentrification and displacement pressure, others suffer from disinvestment and decline. Policy and programmatic strategies are needed to find the right balance in either case. There is a need to invest in diverse communities – strategic public investment, public/private partnership, incentives that stimulate private investment – but at the same time, there is a need to keep things affordable for residents and businesses. If there are zoning and land use restrictions being put in place, their effect on diversity has to be taken into account simultaneously, since such changes often have the affect of driving up rents and driving out poor households (Quigley and Raphael, 2003).

There are many possible policies and programs to choose from to attempt to stabilize and support diversity. A few examples include: strategies to keep units affordable (community land trusts, inclusionary housing requirements); strategies to entice the development of affordable units (tax credits, bonus densities); strategies to preserve neighborhood assets (transfer of development rights, reuse of vacant and tax-foreclosed real estate); strategies to retain rental units (condominium conversion ordinances); strategies to keep buildings occupied (limits on the use of restrictive covenants); strategies to revitalize the public realm (TIFs, loans, grants, bond financing, tax abatement); strategies to distribute tax dollars to the places that most need it (tax-base sharing); and integration maintenance programs, which encourage homeowners to buy property in places where racial/ethnic mix is needed. Organizations like PolicyLink and the Institute for Community Economics keep track of such policies and try to monitor their effects. They have compiled lists of actionable tools focused on promoting social equity in urban revitalization strategies (see especially Harmon, 2003, 2004; and Karlinsky, 2000).

The diverse neighborhoods studied here were often in need of public investment – for parking facilities, for upgrades to schools and parks, for incentives for new businesses or mixed use housing developments. It should be obvious that public investment and the private enterprise it stimulates can result in gentrification and displacement, especially in diverse neighborhoods that have good access to the central city, historic housing stock, pedestrian quality, and lots of potential for revitalization. While this investment has many positive benefits, it also has to be kept from stimulating too much growth. Rapid gentrification is an indicator that the neighborhood is displacing rather than revitalizing (Kennedy and Leonard, 2001). There will need to be policies aimed at retaining rental housing and commercial space for small business, keeping the pace of change slow and steady.

SUSTAINING DIVERSITY THROUGH THE PLANNING PROCESS

There are plenty of ideas about policies and programs to put in place to help diverse places stay diverse – but what of the planning process itself, a process that would presumably precede implementation of policies and programs aimed at preserving diversity? What would a neighborhood planning process devoted to sustaining social diversity be like? There have been many excellent studies of the procedural factors that help in revitalization efforts (e.g., Bright, 2003; Zielenbach, 2000; Keating et al., 1996), but diverse neighborhoods have a different set of issues, requiring a different set of strategies. In the remainder of this chapter, I suggest a planning process specifically geared to the diverse neighborhood.

First and foremost, sustaining diversity at the neighborhood level is going to require neighborhood level planning. Planning at the neighborhood scale has always been a feature of urban planning, although the support of it rises and falls with the availability of municipal funds. Some neighborhood planning approaches have become quite well known. Minneapolis' Neighborhood Revitalization Program has been studied as a unique model of neighborhood empowerment, where each of 66 neighborhoods prepare a neighborhood action plan and request funds for housing, parks, schools, security and the like (Martin and Pentel, 2002; Fainstein and Hirst, 1996). Non-profit groups in Chicago recently established a neighborhood-based planning effort called the New Communities Program, funded by the MacArthur Foundation and LISC (Local Initiative Support Corporation), where neighborhoods are encouraged to develop plans for macro (e.g., job creation) and micro (e.g., block renovation) interventions (McCarron, 2004). As with the Minneapolis neighborhood planning effort, the goal is to bring residents together, build neighborhood capacity among residents, and empower them to make decisions about their neighborhood's future development.

These neighborhood planning efforts are exemplary, but planning for highly diverse areas may require a few extra steps. Neighborhood planning as currently practiced does not really offer a way to account for mixed social structure; it is mostly conceived as being neutral to population heterogeneity vs. homogeneity. There are strategies for getting under-represented people engaged in site planning (Lanfer and Taylor, 2006), and those ideas are useful, but planning for the diverse neighborhood requires a specifically focused approach. In fact the experience of neighborhood planning in diverse areas has shown that there may be a problem getting diverse groups involved. The problem in Chicago was described like this:

> At planning sessions in gentrifying neighborhoods like West Haven on the Near West Side the word 'condo' is spat out like a four-letter word, and yuppies, though key to the location's future, are conspicuous by their absence (McCarron, 2004, p. 12).

The fact that there are no specific neighborhood planning methods developed with diversity in mind is curious given that a vibrant social mix is both a normative goal of planners and a significant challenge to neighborhood stability.

Planning for places currently composed of a complex mix of people requires something beyond inclusive exchange, or a focus only on empowering the under-represented. Because engagement in the planning process always runs the risk of being motivated by a desire for group self-preservation – protecting one group from another in ways that are not mutually reinforcing – something more strategic is required. Those advocating various consensus-building approaches have recognized that dealing with diversity is likely to require something beyond merely democratizing and opening-up public engagement, or simply having a more broad-minded view about the legitimacy of alternative cultural expressions (Innes and Booher, 1999). What needs to be worked out is an approach that enlists ideas about consensus-building and collaborative planning in ways that support socially diverse neighborhoods specifically.

Formulating a planning approach that supports diversity will require an integration of centralized ('top-down') and localized ('bottom-up') approaches. Neighborhood planning is sometimes dichotomized into these two categories: centrally led planning activities that are decentralized at the neighborhood scale, and community-based neighborhood planning, directed by and intended for neighborhood residents. The former is generally written off as too controlling, while the latter is seen as empowering (Checkoway, 1984; Peterman, 1999). But in a diverse neighborhood, both may be required. To begin with, local government planners are needed to help jumpstart the process and help diverse groups come together, since diverse places are prone to apathy and low levels of collective efficacy. In their recent study of diverse neighborhoods in Chicago, Wilson and Taub wrote about the changes that occurred in a neighborhood that had experienced high levels of ethnic in-migration: 'Residents invested little effort in the social organization of the neighborhood ... there was little collective supervision of community activities, and little participation in voluntary organizations such as block clubs, civic and business clubs, parent-teacher organizations, and political groups' (Wilson and Taub, 2006, p. 173). At the same time, neighborhoods with strong community-based capacity and strong resident control have been shown to be the most successful at solving problems (Bright, 2003).

Neighborhood planning for social diversity should embrace difference and work to sustain it. It should do this not by focusing on the empowerment of any particular group, but by ensuring that people are well informed about the diversity at hand, what that diversity means for the future growth of their neighborhood, and the positive ways in which it can be viewed. Tensions and conflicts that arise should not be neutralized or assumed away, but confronted straightforwardly. As much as possible, conflicts that affect the social make-up and well-being of the neighborhood should be put into context and prevented from undermining the diversity of the neighborhood.

Neighborhood planning for diverse areas could, for example, consist of five basic steps:

1   *Target diverse neighborhoods.* The first step would be to decide which neighborhoods should be targeted for a neighborhood planning effort directed at sustaining diversity. This is not the usual procedure. Most often, neighborhoods are selected for planning work based on level of distress or opportunity to stimulate private investment (see, for example, Richmond LISC, 2005, and McCarron, 2004). Taking a somewhat different approach, planners could identify neighborhoods with high levels of social diversity, defined either by income, race, ethnicity, or in relation to some other diversity criteria. Planners could also consider threats to existing diversity and the potential for instability (gentrification, displacement, disinvestment), and the likelihood of success (citizen interest, active and engaged local leadership). Depending on available resources, a number of diverse neighborhoods could be targeted for special planning effort and focus.

2   *Appoint a citizen's planning group.* The second step would be to assemble a citizen's planning group composed of local leaders who represent the diversity of the neighborhood. This group would be formally and strategically recruited, something that research has shown is critical for building citizen participation at the neighborhood level (Marshall, 2004b). The group would be enlisted to support the diversity-sustaining process being proposed. This would be essential given the reality that effective social organization and neighborhood diversity do not generally correlate. As Wilson and Taub (2006) put it, 'strong neighborhoods ... work against the notion of intergroup harmony and integration' (p. 181). One way to counteract that tendency would be to develop a set of shared goals, around which diverse residents could unite. The citizen planning group would be the catalyst for formulating that shared set of objectives.

Planners could help the group work toward an appreciation of what diversity brings to the neighborhood. Rather than being problematized, diversity would be considered as an asset and an essential part of the solution to community problems. Instead of viewing diversity as an imposed condition, whereby income and racial integration is forced by government fiat, diversity would be cast as something positive and unique. A sense of reciprocity and interdependence has to be established, so groups can easily recognize and act on their shared concerns. The group could help articulate, for example, the ways in which the goals of diversity are interrelated: a neighborhood that is open to a range of groups translates to improvements in neighborhood services for all groups. A mix of uses that is good for the economic vitality of a neighborhood adds interest and opportunity for every resident. Diverse people are more likely to have diverse schedules, thereby increasing the ability of the neighborhood to informally patrol its streets at all times of the day. Diversity helps stimulate an expanded set of locally-based social networks, which may be viewed positively by many residents. A community that wants to retain affordability for its children, the workforce, and the elderly is going to have to publicly acknowledge the value of diversity.

3 **Build public awareness.** One of the first tasks for the citizen's planning group would be to look for ways to increase public awareness of neighborhood diversity. Researchers who study diverse neighborhoods have argued that the maintenance of diversity requires 'a publicly stated commitment' to inclusiveness, and that 'image creation and marketing' are important strategies (Maly and Leachman, 1998, p. 154). There needs to be an effort to increase recognition and understanding of the kinds of diversity present, which residents may only have a vague notion of. The ideas to be communicated would need to be simple, straightforward, and visually interesting, presented in a manner that is readily understood, and suitable for publication and exhibition throughout the neighborhood. It should highlight racial, ethnic, income, age and household diversity, and include some explanation about how the level of diversity has changed over time. Graphical output of various kinds could be exhibited in well-traversed public spaces and places, including websites.

The citizens planning group could tap the efforts of local schools. Often, local schools and libraries are involved in projects devoted to celebrating diversity. Children are making art, writing poetry, and finding different means of expression about their cultural identities. They are encouraged to react to their differences with others in positive, affirming ways. Planners should help the citizens planning group spotlight these efforts and integrate them into the neighborhood planning process.

Planners could also help the citizens planning group set up a website that highlights diversity. The website could be used to expose the life stories of a wide range of residents, thereby taking a step toward building familiarity amongst people who otherwise would have no exposure to each other. It is widely recognized that a successful urban neighborhood has a variety of spaces for social and cultural transaction. Planners can make use of technology as one viable 'place' for exchange.

4 **Produce a neighborhood plan.** With a citizen's planning group in place, and the establishment of pre-plan activities focused on building public awareness of neighborhood diversity, the next step would be to formulate a neighborhood plan. A neighborhood plan puts the idea of a shared future in concrete terms. The plan would be used as a framework to channel individual ideas toward something tangible – collectively realized, positive outcomes for the diverse neighborhood. Collaborative planning efforts of this type, often in the form of charrettes, are now recognized as indispensable (Lennertz and Lutzenhiser, 2006). The trick would be to orient them to the specific needs, issues and constraints of the diverse neighborhood.

This is likely to be especially important in a socially diverse neighborhood where lower-valued homes and businesses are seen as a threat to higher-valued ones. One study showed that the neighborhood plan was critical for garnering support for affordable infill housing, because it embedded the infill within a larger context – i.e., the affordable housing was contextualized and planned for (Deitrick and Ellis, 2004). The plan makes it possible to 'envision each building, each development project, in relation to a positive ideal' (Brain, 2005, p. 32). Doing away with negative feelings about subsidized housing, small family-run businesses, or social service agencies in favor of positive feelings about diversity is going to require strong conceptualization of neighborhood, and the plan would be an essential means for accomplishing that.

During the creation of a neighborhood plan in a socially diverse area, it may be especially important to employ participatory methods that avoid 'the tyranny of structurelessness' (Sirianni and Friedland, 2001, p. 24). A variety of methods are possible, from week-long intensive charrettes to months-long visioning efforts. Whatever process is used, the key point for planning would be to keep informed about the trade-offs involved in whatever is being proposed – how the proposals being advocated have consequences for other people in the neighborhood as well as people in the surrounding neighborhoods. If some residents propose, for example, to downzone portions of the neighborhood so that multi-family units are disallowed, or to limit the ability to add accessory units, they need to be made aware of the effect this might have on reducing diversity, the ability to retain essential services they might deem important, the ability to sustain a walkable environment, the increase of traffic arterials through the neighborhood, or the cost of housing for their children, the workforce, and the elderly.

Of course, planners should be well informed about the various design strategies that could be used to stabilize and promote diversity, ideas advocated in the chapters of this book. They should illuminate the importance of housing mix, business mix, facilities that support connection, and the potential benefits of breaking through residential enclaves. They can suggest locations for new non-residential growth – strategically targeted in function of the housing diversity that surrounds them. They could suggest ways to intersperse different housing types, and show how codes could be important for supporting the mix. They could show how strategically placed public investment can support diversity. Planners could be an essential resource on design ideas, pointing out when and where specific ideas being proposed may have the effect of undermining or supporting diversity.

5 *Implement the plan.* Implementation of the neighborhood plan could focus on three things: (1) establishment of a process for shared management of the built environment as an ongoing neighborhood-stabilizing strategy; (2) regulatory reform including new types of codes that encourage a coherent yet flexible guide for the built environment; and (3) recommendations for public investment that stimulates positive change, giving the neighborhood the kinds of improvements it needs without undermining its diversity.

Many studies have shown that implementation of the best laid plans doesn't work if residents aren't directly involved in the process (Halpern, 1995). Going one step further, a neighborhood plan that supports diversity should lay out a process of collaboration whereby the diverse nature of the neighborhood is borne in mind as proposals are presented and responded to. The emerging 'collective intelligence' of recurring responses should not devolve into control by one group or another. There needs to be an operating system that implements whatever shared commitment to place is possible. This could take different forms. A citizen's review committee could be established to provide some oversight of physical change in the neighborhood. Community-based institutions – schools, neighborhood associations, faith-based organizations and the like – could be enlisted to promote participation. Residents in diverse neighborhoods need to be engaged in the shared management of everyday issues.

In fact there are some very creative strategies that have been devised to bring neighborhood residents together to work collectively and implement neighborhood plans. Sometimes these efforts revolve around 'placemaking'. A recent publication by the Project for Public Spaces called *The Great Neighborhood Book: A Do-it-Yourself Guide to Placemaking* (Walljasper, 2007), is full of interesting ideas, like how to create 'places to hang out', how to 'nurture pleasure and pizzazz' and how to organize a neighborhood event in nine simple steps. Strategies like these devoted to bringing residents together to work collectively to strengthen neighborhoods as places are likely to be useful in the effort to stabilize and promote diverse neighborhoods, even if extra effort is likely to be required to get a diverse population involved.

Another appropriate focus of the plan would be to suggest new types of regulatory codes. As argued in this book, diverse neighborhoods need to simultaneously support homeownership and rental housing, integrate a range of housing types, densities, and levels of affordability, and foster a mix of uses, services and facilities. Codes will be essential for making this diversity work. Part of the effort involves getting rid of excessive regulations, like exclusionary zoning and overbearing building codes. Or it involves reversing the rules by which social segregation occurred: allowing multi-family units where they have been excluded, and eliminating minimum lot sizes, maximum densities, minimum setbacks, and other rules that work to prevent housing type diversity. The diversity of uses needed also requires greater flexibility in codes, accompanied by some design control to ensure neighborhood compatibility.

Finally, the neighborhood plan would be implemented through targeted public investment. In particular, diverse neighborhoods need an infrastructure that supports positive social connection, and that means paying particular attention to the public realm. Parks, plazas, streets, and other elements of the public realm sustain diversity by offering shared space as opposed to places of privatized residential space. This stimulates informal, collective control and a sense of shared responsibility. Since diversity decreases distances between elements whose compatibility may be questioned, investment in the public realm may be essential for holding disparate elements together. If residents do not value the external context in which increased proximities occur, instead focusing on the isolated value of the individual dwelling, social diversity may eventually lose support.

Sustaining mixed housing type in a diverse neighborhood is also dependent on strategic investment because diverse neighborhoods are prone to gentrification and displacement pressures. Targeted public investment should be used to ensure that diverse neighborhoods are not permitted to just appreciate beyond the means of middle and lower-income groups. If diverse neighborhoods are also to be well planned and serviced, there will necessarily need to be a political commitment to maintaining their social diversity. This commitment becomes more real if such policies are embedded in the neighborhood plan.

A neighborhood plan with the goal of supporting diversity could also include strategies to increase the viability of mixed-use, mixed-income projects. If presented in the context of a neighborhood plan, diversity-sustaining development projects are likely to engender greater support. Planners could assist residents in laying out strategies for development support, from technical assistance to tax incentives and grant monies, to make it easier for developers and individuals in diverse neighborhoods to combine funding in effective ways. The plan could help ensure a balance between support for larger-scale development and funding for small, independently run businesses.

Above all, planning for neighborhoods that are socially diverse requires a shift in emphasis. There needs to be a focus on stability: how to keep a place diverse and prevent it from being taken over by one particular social group or one particular land use. In such cases, the goal of planning is to encourage change that supports a stable heterogeneity, while discouraging change that undermines it. This will require constant monitoring and strategic thinking. Support of a diverse neighborhood always runs the risk that targeted planning effort will ultimately undermine the very diversity planners and residents had hoped to protect.

CONCLUSION

Design for diversity may be asking the impossible – that urban dwellers use place as a connector rather than a divider. Residents of diverse neighborhoods are being asked to reverse the usual association between place and difference, where attention to one has meant delimiting the other. They are being encouraged to have a heightened sense of place, and at the same time, have a more relaxed attitude about difference.

The strategies of working-class neighborhoods to distance themselves from the ghetto next door are illustrative of this paradox. In Kefalas' study of the working-class 'Beltway' neighborhood in Chicago, residents possessed a strong and distinctive sense of place that tended to separate Beltwayites from residents of poorer neighborhoods nearby. Their strategy for maintaining the health of the neighborhood 'relies on creating social and symbolic distances between themselves and the dispossessed' in an effort to 'deny their own marginality' (Kefalas, 2003, p. 155). Given that the most diverse places in Cook County were often located in neighborhoods adjacent to much poorer ones (Chapter 5), this finding has special relevance.

There is always a danger that designing for social diversity will entail an unacceptable level of control. Many will wonder: How can diversity be 'designed' in any 'authentic' sense? Isn't diversity a product of minimal intervention? (Hough, 1994). And yet, supporting diversity with vague notions about infusing 'discontinuities and inconsistencies' with 'life-affirming opportunities' (Durack, 2002, p. 67) does not sound like much of a strategy. As Kelbaugh writes, such strategies amount to an elevation of the importance of the 'mythic aspect of the ordinary and ugly' (Kelbaugh, 2002, p. 287), not unlike the view that strip malls merely represent a new, as yet under-appreciated, esthetic ideal (Kolb, 2000).

Designing for diversity may be problematic for those bothered by the notion of 'order'. Order has been equated with the attempt to deny social conflict and control the unexpected, whereby humans are organized, but alienated (Boyer, 1983). Planners and designers are viewed as imposers and stiflers who are threatened by the unknown and the uncontrolled. It is a critique legitimately rooted in the fact that many ideas about the spatial planning of cities have been linked to some form of social manipulation (Kostof, 1991).

Yet order is not the enemy of diversity – a point Jane Jacobs cogently made – and I hope that this book has offered a further refinement of this essential theme. Some recognizable sense of order is needed to be able to identify the 'collective' aspects of urbanism. Order is what conveys its public purpose. It allows us to grasp a shared construct, a collective expression that counter-balances the individualism of diversity. Order is what allows diverse urban elements to relate to each other in some way. This implies the need for some control, a point even Jane Jacobs admitted. Left unplanned, profitable uses can start to take over, housing of one kind can begin to dominate, sorting into one primary use can occur, or the scale of building can start to undermine a successful street. Diversity is perfectly capable of self-destructing.

In highly diverse areas, there are special challenges to forming a unified vision of what the neighborhood should be and how it should grow. This makes public participation even more essential, since the ability to take control of neighborhood change may very well be the best strategy for sustaining diversity. Diverse neighborhoods already have to work through social mix on a daily basis. It seems that planners could, at a minimum, ensure that there is a process in place for dealing with conflicts over issues having to do with the design and use of space.

There is evidence that prejudices are declining, and that tolerance for diversity has increased (Farley et al., 1997). And yet, embracing place diversity requires not just a tolerance for diversity, but a tolerance for conflict. It may also require a recreated notion of community. If the basis for commonality is not linked to a common race, ethnicity, social class, occupation, or stage in the life cycle, can it be based on common space? Bella et al. (1996) argued that community is indicative of a withdrawal from social complexity, and admittedly, the ability to translate notions of community beyond nostalgia, conformity and control has been difficult to pull off. But what if community can be based on the overlapping space that diverse groups occupy? And what if planners can help motivate that approach by promoting both the physical design required for diversity and the process needed for its shared management?

Recently, there has been a sense of idealism about diversity that has been reverberating throughout planning scholarship, albeit from differ-ent angles. Dowell Myers (2007), for example, writes about the need to recognize interdependencies and mutual self-interest among the grow-ing 'immigrants and boomers' that will define metropolitan society in the coming decades. Robert Fishman (2005) hopes for the 'reurbanism' of our cities into diverse, mixed-income neighborhoods through a combi-nation of what Jacobs called 'unslumming' and, if need be, a softer form of gentrification. Nan Ellin's (2006) 'Integral Urbanism' is defined by net-works, relationships, connections and interdependencies that counteract separation and retreat. These are part of an emergent, sophisticated dis-course about diversity that goes beyond what has been discussed before.

And yet, we can take comfort in the fact that the quest for diversity has always been part of the planning profession, from Garden Cities to British New Towns to New Urbanism. And there has always been recognition of the significant challenges it entails. It has always been difficult to translate ideals into actual physical form, moving from the rhetoric of 'communities of tolerated difference' to the provision of an actual context for them to grow and flourish.

Harvey Warren Zorbaugh (1929) wrote in the early 20th century that 'the march of the city' was wiping out the 'visible and permanent symbols' that had provided the physical framework for unified communities. Expansion was resulting in the loss of common economic activities and recreational spaces that previously provided a place for 'spontaneous communal activity'. There was no more 'physical basis of local cultural unity' to provide 'unifying and controlling traditions'. Our design challenge now is to see if this physical framework – or some form of it – can be reconstituted. This time, there will be a different twist, one that the Chicago School sociologists may not have thought possible: holding together the community of diversity.

Bibliography

Abrams, Charles, 1955. *Forbidden Neighbors, A Study of Discrimination in Housing.* New York: Harper and Row.

Abu-Lughod, J., Ed., 1994. *From Urban Village to East Village: The Battle for New York's Lower East Side.* Oxford: Blackwell.

Ackerman, Diane, 2004. *An Alchemy of Mind: the Marvel and Mystery of the Brain.* New York: Scribner.

Addams, Jane and Robert Woods et al., 1893. *Philanthropy and Social Progress.* Boston: Thomas Y. Crowell.

Akkerman, Abraham, 2003. Harmonies of urban design and discords of city form. In Alexander R. Cuthbert, Ed., *Designing Cities: Critical Readings in Urban Design.* Oxford: Blackwell, pp. 76–97.

Alba, Richard D., John R. Logan and Brian J. Stults, 2000. The changing neighborhood contexts of the immigrant metropolis. *Social Forces* 79: 587–621.

Alesina, A., R. Baqir and W. Easterly, 1999. Public goods and ethnic divisions. *The Quarterly Journal of Economics.*

Alexander, C., A. Neis, A. Anninou and I. King, 1987. *A New Theory of Urban Design.* New York: Oxford University Press.

Alexander, C., S. Ishikawa, M. Silverstein, M. Jacobson, I. Fiksdahl-King and S. Angel, 1977. *A Pattern Language.* New York: Oxford University Press.

Alexander, Christopher, 1965. A city is not a tree. *Architectural Forum* 122(April): 58–62; 122(May): 58–61.

Allard, Scott W., 2004. *Access to Social Services: The Changing Urban Geography of Poverty and Service Provision.* Washington, DC: The Brookings Institution.

Allen, Chris, Margaret Camina, Rionach Casey, Sarah Coward and Martin Wood, 2005. *Mixed Tenure Twenty Years On: Nothing Out of the Ordinary.* Coventry, UK: Chartered Institute of Housing and the Joseph Rowntree Foundation.

Allen, Peter M., 1999. *Cities and Regions as Self-Organizing Systems: Models of Complexity.* Amsterdam: Gordon and Breach Science Publishers.

Altshuler, Alan, William Morrill, Harold Wolman and Faith Mitchell, Eds., 1999. *Governance and Opportunity in Metropolitan America.* Washington, DC: National Academy Press.

Amin, Ash, 2002. Ethnicity and the multicultural city: Living with diversity. *Environment & Planning A* 34: 959–980.

Anderson, Martin, 1964. *The Federal Bulldozer – A Critical Analysis of Urban Renewal 1949–1962*. Cambridge: The MIT Press.

Andrews, Clinton J., 1999. Putting industrial ecology into place. *Journal of the American Planning Association* 65(4): 364–375.

Angotti, Tom and Eva Hanhardt, 2001. Problems and prospects for healthy mixed-use communities in New York City. *Planning Practice and Research* 16(2): 145–154.

Anonymous, 2003. Chicago most diverse major US economy. *Mortgage Banking* 63(10): 95.

Anselin, Luc, 1988. *Spatial Econometrics: Methods and Models*. Dordrecht: Kluwer Academic Publishers.

Anselin, Luc, 1995. Local Indicators of Spatial Association – LISA. *Geographical Analysis* 27(2): 93–115.

Anselin, Luc, 2002. Under the hood: Issues in the specification and interpretation of spatial regression models. *Agricultural Economics* 17(3): 247–267.

Anselin, Luc and Anil Bera, 1998. Spatial dependence in linear regression models with an introduction to spatial econometrics. In A. Ullah and D. Giles, Eds., *Handbook of Applied Economic Statistics*. New York: Marcel Dekker, pp. 237–289.

Anselin, Luc, I. Syabri and Y. Kho, 2004. GeoDa: An introduction to spatial data analysis. *Geographical Analysis* (forthcoming).

Bacon, Edmund, 1976. *Design of Cities*. New York: Penguin Books.

Banerjee, T., 1993. Antiplanning undercurrents in US planning education: Antithesis or ideology? *Environment and Planning B: Planning and Design* 20: 519–536.

Banerjee, T. and W. C. Baer, 1984. *Beyond the Neighborhood Unit*. New York: Plenum Press.

Barajas, Diego, 2003. *Dispersion: A Study of Global Mobility and the Dynamics of a Fictional Urbanism*. Rotterdam: Episode Publishers.

Barnett, Jonathan, 1995. *The Fractured Metropolis*. New York: Harper Collins.

Barnett, Jonathan, 2006. The way we were, the way we are: The theory and practice of designing cities since 1956. *Harvard Design Magazine* 24(Spring/Summer): 1–4.

Bauer, Catherine, 1951. Social questions in housing and community planning. *Journal of Social Issues* 7: 1–34.

Baumgartner, M. P., 1991. *The Moral Order of a Suburb*. New York: Oxford University Press.

Bayer, Patrick, Robert McMillan and Kim Rueben, 2004. *Residential Segregation in General Equilibrium*. Working paper available online at http://aida.econ.yale.edu/~pjb37/papers.htm

Bayer, Patrick, 2000. *Tiebout Sorting and Discrete Choices: A New Explanation for Socioeconomic Differences in the Consumption of School Quality*. Working paper available online at http://aida.econ.yale.edu/~pjb37/papers.htm

Beard, Victoria A., 2003. Learning radical planning: The power of collective action. *Planning Theory* 2(1): 13–35.

Beatley, Timothy and Kristy Manning, 1997. *The Ecology of Place: Planning for Environment, Economy, and Community*. Washington, DC: Island Press.

Bell, Derrick, 2004. *Silent Covenants: Brown v. Board of Education and the Unfulfilled Hopes for Racial Reform*. New York: Oxford University Press.

Bellah, Robert N., Richard Madsen, William M. Sullivan, Ann Swidler and Steven M. Tipton, 1996. *Habits of the Heart: Individualism and Commitment in American Life*. Berkeley: University of California Press.

Berry, Brian J. L., 2001. *A new urban ecology? Urban Geography* 22(8): 699–701.

Berube, Alan and Thacher Tiffany, 2004. *The Shape of the Curve: Household Income Distributions in US Cities, 1979–1999*. Washington, DC: The Brookings Institution.

Berube, Alan and Elizabeth Kneebone, 2006. *Two Steps Back: City and Suburban Poverty Trends 1999–2005*. Washington, DC: The Brookings Institution. http://www.brookings.edu/metro/pubs/20061205_citysuburban.htm

Birch, Eugenie Ladner, 2002. Having a Longer View on Downtown Living. *Journal of the American Planning Association* 68(1): 5–21.

Blakely, E. J. and M. G. Snyder, 1995. *Fortress America: Gated and Walled Communities in the United States*. Cambridge, MA: Lincoln Institute of Land Policy/Brookings Institution.

Boal, Frederick W., 2001. *Urban Ethnic Segregation and the Scenarios Spectrum*. Paper presented at the International Seminar on Segregation in the City, July 26–28, 2001, at the Lincoln Institute of Land Policy.

Bohl, Charles, 2000. New Urbanism and the city: potential applications and implications for distressed inner-city neighborhoods. *Housing Policy Debate* 11(4): 761–801.

Bollens, Scott, 1999. *Urban Peace-Building in Divided Societies: Belfast and Johannesburg*. Boulder, CO: Westview Press.

Booher, David E. and Judith E. Innes, 2002. Network power in collaborative planning. *Journal of the American Planning Association* 21: 221–236.

Borgman, Albert, 1987. *Technology and the Character of Contemporary Life*. Chicago: University of Chicago Press.

Bourne, Larry S. and David F. Ley, Eds., 2002. *Changing Social Geography of Canadian Cities*. Montreal, Canada: McGill-Queen's University Press.

Bowden, Charles and Lew Kreinberg, 1981. *Street Signs Chicago: Neighborhood and Other Illusions of Big-City Life*. Chicago: Chicago Review Press.

Boyer, M. Christine, 1983. *Dreaming the Rational City: The Myth of American City Planning*. Cambridge: MIT Press, p. 7.

Boyer, M. Christine, 1990. Erected against the city: The contemporary discourses of architecture and planning. *Center 2: Architecture vs. Planning*: 36–43.

Brain, David, 2005. From Good Neighborhoods to Sustainable Cities: Social Science and the Social Agenda of the New Urbanism. *International Regional Science Review* 28: 217–238.

Brandt, J. and H. Vejre, 2004. *Multifunctional Landscapes Volume I: Theory, Values and History*. Boston, MA: MIT Press.

Brasington, David M., 2002. Differences in the production of education across regions and urban and rural areas. *Regional Studies* 36(2): 137–146.

Breslav, M. A., A. R. Berkowitz, C. H. Nilon and K. S. Hollweg, 2000. Cities are ecosystems! New trend to study urban areas. *Ecological Economics* 32: 337–339.

Briggs, Xavier de Souza, Ed., 1998. Racially and Ethnically Diverse Urban Neighborhoods. *Cityscape: A Journal of Policy Development and Research* 4(2).

Briggs, Xavier de Souza, Joe T. Darden and Angela Aidala, 1999. In the wake of desegregation: Early impacts of scattered-site public housing on neighborhoods in Yonkers, New York. *Journal of the American Planning Association* 65(1): 27–49.

Briggs, Xavier de Souza, 2001. *Ties that Bind, Bridge and Constrain: Social Capital and Segregation in the American Metropolis.* Paper presented at the International Seminar on Segregation in the City, July 26–28, 2001, at the Lincoln Institute of Land Policy.

Briggs, Xavier de Souza, 2005. *The Geography of Opportunity: Race and Housing Choice in Metropolitan America.* Washington, DC: Brookings Institution Press.

Briggs, Xavier de Souza, Elizabeth J. Mueller and Mercer L. Sullivan, 1997. *From Neighborhood to Community: Evidence on the Social Effects of Community Development.* New York: New School for Social Research.

Bright, Elise M., 2003. *Reviving America's Forgotten Neighborhoods.* London: Routledge.

Brooke, Nicholas, 2003. Focus on global issues: What makes for a successful city. *Real Estate Issues* 28(4): 28–29.

Brooks, David, 2004. Our sprawling, supersize utopia. *The New York Times* April 4.

Brophy, Paul C. and Rhonda N. Smith, 1997. Mixed-income housing: Factors for success. *Citiscape: A Journal of Policy Development and Research* 3(2): 3–31.

Brown, Barbara B., Douglas D. Perkins and Graham Brown, 2004. Crime, new housing, and housing incivilities in a first-ring suburb: Multilevel relationships across time. *Housing Policy Debate* 15(2): 301.

Bruegmann, Robert, 2005. *Sprawl: A Compact History.* Chicago: University of Chicago Press.

Bullard, Robert D., Glenn S. Johnson and Angel O. Torres, 2000. *Sprawl City: Race, Politics and Planning in Atlanta.* Washington, DC: Island Press.

Burayidi, Michael A., Ed., 2000. *Urban Planning in a Multicultural Society.* Westport, CN: Praeger.

Burgess, E., 1925. The growth of a city. In R. Park, E. Burgess and R. McKenzie, Eds., *The City.* Chicago: University of Chicago Press.

Burtless, Gary, 1996. *Does Money Matter: The Effects of School Resources on Student Achievement and Adult Success.* Washington, DC: Brookings Institution.

Burton, Elizabeth, 2002. Measuring urban compactness in UK towns and cities. *Environment and Planning B: Planning and Design* 29: 219–250.

Burton, Elizabeth and Lynne Mitchell, 2006. *Inclusive Urban Design: Streets for Life.* Oxford: Architectural Press.

Byrne, John and John Flaherty, 2004. *Measuring Diversity in Australian Residential Property.* Paper presented at the Pacific Rim Real Estate Society, Tenth Annual Conference, January 25–28. Bangkok, Thailand: Thammasat University.

Calavita, Nico and Kenneth Grimes, 1998. Inclusionary zoning in California: The experience of two decades. *Journal of the American Planning Association* 64(2): 150–169.

Calow Peter, Ed., 1998. *The Encyclopedia of Ecology & Environmental Management.* Oxford: Blackwell Science.

Calthorpe, Peter, 1993. *The Next American Metropolis: Ecology, Community, and the American Dream.* Princeton, NJ: Princeton Architectural Press.

Capra, Fritjof, 1996. *The Web of Life: A New Scientific Understanding of Living Systems.* New York: Anchor Books.

Carmon, Naomi, 2002. The Phoenix Strategy for updating housing stock: Preventing neighborhood deterioration and promoting sustainable development. *Journal of the American Planning Association* 68(4): 416–434.

Caro, Robert A., 1974. *The Power Broker: Robert Moses and the Fall of New York.* New York: Alfred A. Knopf.

Cashin, Sheryll, 2004. *The Failures of Integration: How Race and Class are Undermining the American Dream.* New York: Public Affairs.

Castells, Manuel, 1983. *The City and the Grassroots.* Berkeley: University of California Press.

Castells, Manuel, 2003. The new historical relationship between space and society. In Alexander R. Cuthbert, Ed., *Designing Cities: Critical Readings in Urban Design.* Oxford: Blackwell, pp. 59–68.

Caulfield, J., 1994. *City Form and Everyday Life: Toronto's Gentrification and Critical Social Practice.* Toronto: University of Toronto Press.

Cervero, Robert, 1996. Mixed land-uses and communting: Evidence from the American housing survey. *Transportation Research* 30(5): 361–377.

Chapman, Nancy J. and Deborah A. Howe, 2001. Accessory apartments: Are they a realistic alternative for ageing in place? *Housing Studies* 16(5): 637–651.

Checkoway, Barry, 1984. Two types of planning in neighborhoods. *Journal of Planning Education and Research* 3: 102–109.

Chichilnisky, Graciela, 1994. *The American Economic Review* 84(2): 427.

Clampet-Lundquist, Susan, 2004. HOPE VI relocation: Moving to new neighborhoods and building new ties. *Housing Policy Debate* 15(2): 415–447.

Clark, W. A. V., 2001. Residential segregation trends. In Stephen Thernstrom and Abigail Thernstrom, Eds., *Beyond the Color Line.* Stanford: Hoover Institute Press.

Clark, William A. V., 1998. *The California Cauldron: Immigration and the Fortunes of Local Communities.* Guilford Press.

Clark, William A. V., 1993. Neighborhood transitions in multiethnic/racial contexts. *Journal of Urban Affairs* 15(2): 161–172.

Clark, W. A. V. and F. M. Dieleman, 1996. *Households and Housing: Choice and Outcomes in the Housing Market.* New Brunswick, NJ: Rutgers University, Center for Urban Policy Research.

Clay, Grady, 1994. *Real Places: An Unconventional Guide to America's Generic Landscape.* Chicago: University of Chicago Press.

Cohen, James R., 1998. Combining Historic Preservation and Income Class Integration: A Case Study of the Butchers Hill Neighborhood of Baltimore. *Housing Policy Debate* 9(3): 663–697.

Cole, Ian and Barry Goodchild, 2001. Social mix and the 'balanced community' in British housing policy – a tale of two epochs. *GeoJournal* 51: 351–360.

Collins, James *et al.*, 2000. The new urban ecology. *American Scientist* 88: 5.

Congress for the New Urbanism, 2000. *Charter of the New Urbanism*. New York: McGraw-Hill.

Congress for the New Urbanism, 2004. *Codifying New Urbanism: How to Reform Municipal Land Development Regulations*. Washington, DC: American Planning Association.

Coulton, Claudia J., 2003. Metropolitan inequities and the ecology of work: Implications for welfare reform. *Social Service Review* 77: 159.

Crompton, Andrew, 2001. The fractal nature of the everyday environment. *Environment and Planning B: Planning and Design* 28: 243–254.

Crump, J., 2002. Deconcentration by demolition: public housing, poverty, and urban policy. *Environment and Planning D: Society and Space* 20: 581–596.

Cullen, Gordon, 1961. *Townscape*. London: Architectural Press.

Culp, Lisa-Anne, 2003. Mixed results for mixed-use. *National Real Estate Investor* August 1 2003.

Cuthbert, Alexander R., 2006. *The Form of Cities: Political Economy and Urban Design*. Oxford: Blackwell.

Cutler, David M. and Edward L. Glaeser, 1997. Are ghettos good or bad? *Quarterly Journal of Economics* 112(3): 827–872.

Davis, Mike, 1990. *City of Quartz: Excavating the Future of Los Angeles*. London: Verso.

Davis, Mike, 2005. Ozzie and Harriet in hell: On the decline of inner suburbs. In William S. Saunders, Ed., *Sprawl and Suburbia*. Minneapolis: University of Minnesota Press, pp. 27–33.

Dawkins, Casey J., 2004. Measuring the spatial pattern of residential segregation. *Urban Studies* 41(4): 833–851.

Day, Kristen, 2003. New Urbanism and the Challenge of Designing for Diversity. *Journal of Planning Education and Research* 23(1): 83–95.

Deitrick, Sabina and Cliff Ellis, 2004. New Urbanism in the inner city: A case study of Pittsburgh. *Journal of the American Planning Association* 70(4): 426–443.

Delany, Samuel R., 1999. Three, Two, One, Contact: Times Square Red, 1998. In Joan Copjec and Michael Sorkin, Eds., *Giving Ground: The Politics of Propinquity*. London: Verso, pp. 19–85.

Ding, Chengri and Gerrit Knaap, 2002. Property values in inner-city neighborhoods: the effects of homeownership, housing investment, and economic development. *Housing Policy Debate* 13(4): 701–727.

Dolores, Hayden, 2003. *Building Suburbia*. New York: Pantheon Books, p. 121.

Down, Anthony, 1991. The advisory commission on regulatory barriers to affordable housing: Its behavior and accomplishments. *Housing Policy Debate* 2(4): 1095–1137.

Downs, Anthony, 1994. *New Visions for Metropolitan America.* Washington, DC: The Brookings Institution.

Downs, Anthony, 1999. Some realities about sprawl and urban decline. *Housing Policy Debate* 10(4): 955–994.

Downs, Anthony, 1999. *Some realities about sprawl and urban decline.* Washington, DC: The Brookings Institution.

Downs, Anthony, 2000. How city planning practices affect metropolitan-area housing markets, and vice versa. In L. Rodwin and B. Sanyal, Eds., *The Profession of City Planning.* New Brunswick: Center for Urban Policy Research.

Dramstad, Wenche E., James D. Olson and Richard T. T. Forman, 1996. *Landscape Ecology Principles in Landscape Architecture and Land-Use Planning.* Cambridge, MA and Washington, DC: Harvard University and Island Press.

Dreier, Peter, John Mollenkopf and Todd Swanstrom, 2001. *Place Matters: Metropolitics for the Twenty-First Century.* Lawrence, KS: University Press of Kansas.

Duany, Andres and Emily Talen, 2002. Transect planning. *Journal of the American Planning Association* 68(3): 245–266.

Duany, Andres, Elizabeth Plater-Zyberk and Jeff Speck, 2000. *Suburban Nation: The Rise of Sprawl and the Decline of the American Dream.* New York: North Point Press.

Duany, Andres, 2004. The Celebration controversies. *International Network for Traditional Building, Architecture & Urbanism* 1(8). On-line publication available at www.intbau.org.

Duany, Andres, Elizabeth Plater-Zyberk and Robert Alminana, 2003. *The New Civic Art: Elements of Town Planning.* New York: Rizzoli.

Duncan, James S. and Nancy G. Duncan, 2004. *Landscapes of Privilege: The Politics of the Aesthetic in an American Suburb.* New York: Routledge.

Dunkley, Bill, Amy Helling and David S. Sawicki, 2004. Accessibility versus scale: Examining the tradeoffs in grocery stores. *Journal of the American Planning Association* 23(4): 387–401.

Durack, Ruth, 2002. Village Vices: The contradiction of new urbanism and sustainability. *Places* 14(2): 64–69, p. 67.

Duranton, G. and D. Puga, 2001. Nursery cities: Urban diversity, process innovation, and the life cycle of products. *American Economic Review* 91(5): 1454–1477.

Ellen, Ingrid Gould and Margery Austin Turner, 1997. Does neighborhood matter? Assessing recent evidence. *Housing Policy Debate* 8(4): 833–866.

Ellen, Ingrid Gould, 1998. Stable racial integration in the contemporary United States: An empirical overview. *Journal of Urban Affairs* 20(1): 27–42.

Ellen, Ingrid Gould, 2000a. Race-based neighborhood projection: a proposed framework for understanding new data on racial integration. *Urban Studies* 37(9): 1513–1533.

Ellen, Ingrid Gould, 2000b. *Sharing America's Neighborhoods: The Prospects for Stable, Racial Integration.* Cambridge, MA: Harvard University Press.

Ellin, Nan, 1996. *Postmodern Urbanism*. New York: Princeton Architectural Press, p. 157.

Ellin, Nan, 1997. *Architecture of Fear*. Princeton: Princeton Architectural Press.

Ellin, Nan, 2006. *Integral Urbanism*. New York: Routledge.

Engwicht, David, 2003. *Creating Creative Cities. Creative Communities International.* Available on-line at http://www.lesstraffic.com/Articles/Place/creative.htm.

Espino, N. Ariel, 2001. *Inequality, Segregation and Housing Markets: A Theoretical Overview and a Qualitative Profile of the US Case*. Paper presented at the International Seminar on Segregation in the City, July 26–28, 2001, at the Lincoln Institute of Land Policy.

Evans, Alan, 1974. Economic influences on social mix. *Urban Studies* 13: 247–260.

Ewing, Reid, Rolf Pendall and Don Chen, 2003. *Measuring Sprawl and Its Impact*. Washington, DC: Smart Growth America. http://www.smartgrowthamerica. org/sprawlindex/sprawlindex.html

Ewing, Reid, 1996. *Best Development Practices*. Washington, DC: Planners Press.

Fainstein, Susan, 2000. New directions in planning theory. *Urban Affairs Review* 35(4): 451–478.

Fainstein, Susan, Ian Gordan and Michael Harole, 2000. *Divided Cities: New York and London in the Contemporary World*. Oxford: Blackwell.

Fainstein, Susan S. and Clifford Hirst, 1996. Neighborhood organizations and community planning: The Minneapolis Neighborhood Revitalization Program. In W. Dennis Keating, Norman Krumholz and Philip Star, Eds., *Revitalizing Urban Neighborhoods*. Lawrence, KS: University Press of Kansas, pp. 96–111.

Farley, R., E. L. Fielding and M. Krysan, 1997. The residential preferences of Blacks and Whites: A four-metropolis analysis. *Housing Policy Debate* 8: 763–800.

Farley, Reynolds and William H. Frey, 1994. Changes in the segregation of whites from blacks during the 1980s: small steps toward a more integrated society. *American Sociological Review* 59(1): 23–45.

Fasenfest, David, Jason Booza and Kurt Metzger, 2004. *Living Together: A New Look at Racial and Ethnic Integration in Metropolitan Neighborhoods, 1990– 2000*. Washington, DC: The Brookings Institution.

Fauth, Rebecca C., 2004. The impacts of neighborhood poverty deconcentration efforts on low-income children's and adolescents' well-being. *Children, Youth and Environments* 14(1).

Feins, Judith D. and Mark D. Shroder, 2005. Moving to opportunity: The demonstration's design and its effects on mobility. *Urban Studies* 42(8): 1275–1299.

Feldman, Roberta and Susan Stall, 2004. *The Dignity of Resistance: Women Residents' Activism in Chicago Public Housing*. Cambridge University Press.

Filion, Pierre and Karen Hammond, 2003. Neighbourhood land use and performance: The evolution of neighbourhood morphology over the 20th century. *Environment and Planning B: Planning and Design* 30: 271–296.

Filion, Pierre, Kathleen McSpurren and Nancy Huether, 2000. Synergy and movement within suburban mixed-use centers: The Toronto experience. *Journal of Urban Affairs* 22(4): 419–438.

Filion, Pierre, Trudi Bunting and Keith Warriner, 1999. The entrenchment of urban dispersion: residential preferences and location patterns in the dispersed city. *Urban Studies* 36(8): 1317–1347.

Fincher, Ruth and Jane M. Jacobs, 1998. *Cities of Difference*. New York: Guilford Press.

Finkel, Meryl and Larry Buron, 2001. *Study on Section 8 Voucher Success Rates*. Washington, DC: US Department of Housing and Urban Development.

Fischer, Claude S., Gretchen Stockmayer, Jon Stiles and Hout Michael, 2004. Distinguishing the Geographic Levels and Social Dimensions of US Metropolitan Segregation, 1960–2000. *Demography* 41(1): 37–59.

Fischer, Claude S., 1982. *To Dwell Among Friends: Personal Networks in City and Town*. Chicago: University of Chicago Press.

Fischer, Claude S., 1975. Toward a subcultural theory of urbanism. *American Journal of Sociology* 80(6): 1319–1341.

Fishman, Robert, 2000. The metropolitan tradition in American planning. In Robert Fishman, Ed., *The American Planning Tradition*. Washington, DC: Woodrow Wilson Center Press, pp. 65–85.

Fishman, Robert, 2002. The bounded city. In Kermit C. Parsons and David Schuyler, Eds., *From Garden City to Green City*. Baltimore: Johns Hopkins University Press, pp. 58–66.

Fishman, Robert, 2004. Rethinking public housing. *Places, a Forum of Environmental Design* 16(2): 26–33.

Fishman, Robert, 2005. The Fifth Migration. *Journal of the American Planning Association* 71, 4: 357–366.

Florida, Richard, Gary Gates, Brian Knudsen and Kevin Stolarick, 2003. *Beyond Spillovers: The Effects of Creative-density on Innovation*. Working paper, Carnegie Mellon University, H. John Heinz III School of Public Policy and Management, Pittsburgh, PA.

Florida, Richard, 2002a. *The Rise of the Creative Class*. New York: Basic Books.

Florida, Richard, 2002b. The economic geography of talent. *Annals of the Association of American Geographers* 92(4): 743–755.

Florida, Richard, 2004. Revenge of the Squelchers. *The Next American City*. Available on-line at http://www.americancity.org.

Fogelson, Robert, 2005. *Bourgeois Nightmares: Suburbia, 1870–1930*. New Haven: Yale University Press.

Forest, B., 2002. Hidden segregation? The limits of geographically based affirmative action. *Political Geography* 21: 855–880.

Forsyth, Ann and Laura Musacchio, 2005. *Designing for Small Parks: A Manual for Addressing Social and Ecological Concerns*. New York: Wiley.

Frank, Lawrence D., Peter O. Engelke and Thomas L. Schmid, 2006. *Health and Community Design: The Impact of the Built Environment on Physical Activity*. Washington, DC: Island Press.

Frankenberg, E. and C. Lee, 2003. Charter Schools and Race: A Lost Opportunity for Integrated Education. *American Educational Research Association* 11(32).

Frankenberg, E., C. Lee and G. Orfield, 2003. *A multiracial society with segregated schools: Are we losing the dream?* Cambridge, MA: The Civil Rights Project at Harvard University.

Freedman, Samuel G., 2004. Still separate, still unequal. *The New York Times* May 16.

Freeman, Lance, 2005. *There Goes the 'Hood': Views of Gentrification from the Ground Up.* Philadelphia: Temple University Press.

Freeman, Lance and Frank Braconi, 2004. Gentrification and displacement: New York City in the 1990s. *Journal of the American Planning Association* 70(1): 39–53.

Frey, William H., 2001. "Melting Pot Suburbs: A Census 2000 Study of Suburban Diversity" Census 2000 Series, Center on Urban and Metropolitan Studies. Washington, DC: The Brookings Institution.

Frieden, Bernard and Marshall Kaplan, 1975. *The Politics of Neglect.* Erkeley: University of California Press.

Friedmann, John, 2002. City of fear or open city? *Journal of the American Planning Association* 68(3): 237–243.

Gallagher, Winifred, 1994. *The Power of Place: How Our Surroundings Shape Our Thoughts, Emotions, Actions.* New York: Perennial.

Galster, George C., 2005. *Low-Income Households in Mixed-Income Neighborhoods: Extent, Trends, and Determinants.* Report to US Department of Housing and Urban Development, August, 2005 (with J. Booza, K. Metzger & J. Cutsinger).

Galster, George C., 1990. Neighborhood racial change, segregationist sentiments, and affirmative marketing policies. *Journal of Urban Economics* 27: 334–361.

Galster, George C. and Edward W. Hill, 1992. *The Metropolis in Black and White: Place, Power and Polarization.* New Brunswick, NJ: Center for Urban Policy Research.

Galster, George C. and Donald DeMarco, 1993. Pro-integrative policy: Theory and practice. *Journal of Urban Affairs* 15(2): 141–160.

Galster, George C. and Sean P. Killen, 1995. The geography of metropolitan opportunity: A reconnaissance and conceptual framework. *Housing Policy Debate* 6(1): 7–44.

Galster, George C., Peter A. Tatian, Anna M. Santiago, Kathryn L. S. Pettit and Robin E. Smith, 2003. *Why Not in My Backyard? Neighborhood Impacts of Deconcentrating Assisted Housing.* New Brunswick, NJ: Center for Urban Policy Research.

Galster, George, 1998. A stock/flow model of defining racially integrated neighborhoods. *Journal of Urban Affairs* 20(1): 43–51.

Galster, G. C., J. Booza, K. Metzger and J. Cutsinger, 2005. Low-income households in mixed-income neighborhoods: Extent, trends, and determinants. Washington, DC: Report to US Department of Housing and Urban Development, (August).

Gans, Herbert, 1961. The balanced community: Homogeneity or heterogeneity in residential areas? *American Institute of Planners Journal* 27(3): 176–184.

Garvin, Alexander, 2002. *The American City: What Works, What Doesn't.* New York: McGraw-Hill.

Geddes, Patrick, 1915. *Cities in Evolution*. London: Williams & Norgate.

Gehl, J., 1987. *Life Between Buildings: Using Public Space*. New York: Van Nostrand Reinhold.

Gerckens, Laurence C., 1994. American zoning and the physical isolation of uses. *Planning Commissioners Journal* 15: 10.

Gilroy, Rose and Chris Booth, 1999. Building an infrastructure of everyday lives. *European Planning Studies* 7(3): 307–325.

Girling, Cynthia and Ronald Kellett, 2005. *Skinny Streets and Green Neighborhoods: Design for Environment and Community*. Washington: Island Press.

Glaeser, Edward L. and Jacob Vigdor, 2003. Racial segregation: Promising news. In Bruce Katz and Robert E. Lang, Eds., *Redefining Urban and Suburban America, Evidence from Census 2000*. Washington, DC: Brookings Institution Press.

Glaeser, Edward, H. D. Kallal, J. A. Scheinkman and A. Shleifer, 1992. Growth in cities. *Journal of Political Economy* 100(6): 1127–1152.

Glaeser, Edward, 2000. The future of urban research: Nonmarket interactions. *Brookings-Wharton Papers on Urban Affairs*: 101–150.

Glazer, Nathan and Daniel P. Moynihan, 1963. *Beyond the Melting Pot: The Negroes, Puerto Ricans, Jews, Italians, and Irish of New York City*. Cambridge, MA: MIT Press.

Glazer, Nathan, 1959. The school as an instrument in planning. *American Institute of Planners Journal* 25(4): 191–199.

Gleeson, Brendan and Neil Sipe, Eds., 2006. *Creating Child Friendly Cities: Reinstating Kids in the City*. London: Routledge.

Gober, P., K. E. McHugh and N. Reid, 1991. Phoenix in flux: Household instability, residential mobility and neighborhood change. *Annals, Association of American Geographers* 81.

Goering, John and Judith D. Feins, 2003. *Choosing a Better Life? Evaluating the Moving to Opportunity Social Experiment*. Washington, DC: Urban Institute Press.

Goetz, Edward G., 1996. *Clearing the Way: Deconcentrating the Poor in Urban America*. Washington, DC: Urban Institute Press.

Goldsmith, Stephen, 2002. *Putting Faith in Neighborhoods: Making Cities Work through Grassroots Citizenship*. Washington, DC: Hudson Institute.

Gosling, David, 2003. *The Evolution of American Urban Design*. West Sussex, UK: Wiley-Academy.

Gottdiener, Mark and Joe R. Feagin, 1998. The paradigm shift in urban sociology. *Urban Affairs Quarterly* 24: 163–187.

Grannis, Rick, 2003. T-Communities: Pedestrian Street Networks and Residential Segregation in Chicago, Los Angeles, and New York. Working paper.

Granovetter, Mark S., 1983. The Strength of Weak Ties: A Network Theory Revisited. *Sociological Theory* 1: 201–233.

Granovetter, Mark S., 1990. The Old and the New Economic Sociology: A History and an Agenda. In R. Friedland and A. F. Robertson, Eds., *Beyond the Marketplace: Rethinking Economy and Society*. New York: Aldine, pp. 89–112.

Grant, Jill, 2002. Mixed use in theory and practice: Canadian experience with implementing a planning principle. *Journal of the American Planning Association* 68(1): 71–84.

Greed, Clara and Marion Roberts, Eds., 1998. *Introducing Urban Design: Interventions and Responses*. Essex, UK: Longman.

Greenbaum, Susan and Paul Greenbaum, 1985. The ecology of social networks in four urban neighborhoods. *Social Networks* 7: 47–76.

Greenberg, Joseph H., 1956. The measurement of linguistic diversity. *Language* 32: 109–115.

Greenberg, Mike, 1995. *The Poetics of Cities: Designing Neighborhoods that Work*. Columbus, OH: Ohio State University Press.

Greenbie, Barrie B., 1978. A model for accommodating the human need for small-scale communities within the context of global cooperative systems. *Urban Ecology* 3: 137–153.

Grigsby, W., M. Baratz, G. Galster and D. MacLennan, 1987. *The Dynamics of Neighbourhood Change and Decline*. Oxford: Pergamon.

Grossman, James R., Ann Durkin Keating and Janice L. Reiff, Eds., 2004. *The Encyclopedia of Chicago*. Chicago: University of Chicago Press.

Grusky, David B. and Szonja Szelenyi, Eds., 2006. *Inequality: Classic Readings in Race, Class and Gender*. Boulder, CA: Westview Press.

Hall, Peter, 2002. *Cities of Tomorrow: An Intellectual History of Urban Planning and Design in the Twentieth Century*, 3rd edn. Oxford: Blackwell.

Halpern, Robert, 1995. *Rebuilding the Inner City: A History of Neighborhood Initiatives to Address Poverty in the United States*. New York: Columbia University Press.

Handy, Susan, 1996. Understanding the link between urban form and nonwork travel behavior. *Journal of Planning Education and Research* 15: 183–198.

Hanhorster, Heike, 2000. Whose neighbourhood is it? Ethnic diversity in urban spaces in Germany. *GeoJournal* 51(4): 329–338.

Harmon, Tasha, 2003. *Integrating Social Equity and Growth Management*. Springfield, MA: The Institute for Community Economics.

Harmon, Tasha, 2004. *Integrating Social Equity and Smart Growth*. Springfield, MA: The Institute for Community Economics.

Harwood, Stacy Anne, 2005. Struggling to Embrace Difference in Land-Use Decision Making in Multicultural Communities. *Planning, Practice and Research* 20(4): 355–371.

Hartman, Craig W., 2000. Memory palace, place of refuge, Coney Island of the mind: The evolving roles of the library in the late 20th century. *Research Strategies* 17: 107–121.

Harvey, David, 1989. *The Condition of Postmodernity*. Oxford: Basil Blackwell.

Harvey, David, 1997. The New Urbanism and the communitarian trap. *Harvard Design Magazine* Winter/Spring: 68–69.

Harvey, David, 2000. *Spaces of Hope*. Berkeley: University of California Press.

Hawley, Amos H., 1950. *Human Ecology: A Theory of Community Structure*. New York: Ronald Press.

Hayden, Dolores, 1980. What would a non-sexist city be like? Speculations on housing, urban design, and human work. *Signs* 5/3(supplement): s181.

Hayden, Dolores, 1986. *Redesigning the American Dream.* New York: W.W. Norton.

Hayden, Dolores, 2003. *Building Suburbia: Green Fields and Urban Growth, 1820–2000.* New York: Pantheon Books.

Healey, Patsy, 1997. *Collaborative Planning – Shaping Places in Fragmented Societies.* London: MacMillan Press.

Heclo, Hugh, 1994. Poverty politics. In Sheldon Danziger, Gary Sandefur and Daniel Weinberg, Eds., *Confronting Poverty: Prescriptions for Change.* New York: Russell Sage Foundation, pp. 396–437.

Hegemann, Werner and Elbert Peets, 1922. *The American Vitruvius: An Architects' Handbook of Civic Art.* New York: The Architectural Book Publishing Co., Paul Wenzel & Maurice Krakow.

Hillier, B. and J. Hanson, 1984. *The Social Logic of Space.* Cambridge: Cambridge University Press.

Hise, Greg, 2001. 'Nature's workshop' industry and urban expansion in Southern California, 1900–1950. *Journal of Historical Geography* 27(1): 74–92.

Hise, Greg, 2004. *Mexicans and that Sort of Thing: Race and Social Distance in Los Angeles.* Paper presented at the symposium Constructing Race: The Built Environment, Minoritization, and Racism in the United States. University of Illinois, Urbana-Champaign, March 5–6.

Holcomb, Briavel, 1986. Review of *Gentrification, Displacement and Neighborhood Revitalization,* J. John Palen and Bruce London, Eds. *Professional Geographer,* p. 441.

Hough, Michael, 1994. Place-making and design review. In Brenda Case Scheer and Wolfgang F. E. Preiser, Eds., *Design Review: Challenging Urban Aesthetic Control.* New York: Chapman & Hall, pp. 147–155.

Hudnut III, William H., 2003. *Halfway to Everywhere: A Portrait of America's First-Tier Suburbs.* Washington, DC: The Urban Land Institute.

Hughes, James W. and Joseph J. Seneca, 2004. *The Beginning of the End of Sprawl?* Rutgers Regional Report, Issue Paper No. 21. New Brunswick, NJ: Edward J. Bloustein School of Planning and Public Policy.

Huie, Stephanie Bond and W. Parker Frisbie, 2000. The Components of Density and the Dimensions of Residential Segregation. *Population Research and Policy Review* 19: 505–524.

Hurlbert, S., 1971. The non-concept of species diversity: A critique and alternative parameters. *Ecology* 52: 577–586.

Ihlanfeldt, Keith R., 2004. Introduction: Exclusionary land-use regulations. *Urban Studies* 41(2): 255–260.

Ihlanfeldt, Keith R. and D. L. Sjoquist, 1998. The spatial mismatch hypothesis: A review of recent studies and their implications for welfare reform. *Housing Policy Debate* 9(4): 849–892.

Ihlanfeldt, Keith R. and Benjamin P. Scafidi, 2002. The neighborhood contact hypothesis: Evidence from a multicity study of urban inequality. *Urban Studies* 39(4): 619–641.

Immergluck, Dan and Geoff Smith, 2003. Measuring neighborhood diversity and stability in home-buying: Examining patterns by race and income in a robust housing market. *Journal of Urban Affairs* 25(4): 473–491.

Innes, Judith E. and David E. Booher, 1999. Consensus Building as Role Playing and Bricolage: Toward a Theory of Collaborative Planning. *Journal of the American Planning Association* 65(1): 9–26.

Jackson, Kenneth T., 1985. *Crabgrass Frontier: The Suburbanization of the United States*. Oxford: Oxford University Press.

Jacobs, Allan B., 1993. *Great Streets*. Cambridge, MA: MIT Press.

Jacobs, Allan and Donald Appleyard, 1987. Toward an urban design manifesto. *Journal of the American Planning Association* 53: 1.

Jacobs, Jane, 1961. *The Death and Life of Great American Cities*. New York: Vintage Books.

Jargowsky, Paul A., 1997. *Poverty and Place: Ghettos, Barrios, and the American City*. New York: Russell Sage Foundation.

Jargowsky, Paul A., 2002. Sprawl, Concentration of Poverty, and Urban Inequality. In Gregory D. Squires, Ed., *Urban Sprawl: Causes, Consequences & Policy Responses*. Washington, DC: The Urban Institute Press, pp. 39–71.

Jargowsky, Paul A., 2003. *Stunning Progress, Hidden Problems: The Dramatic Decline of Concentrated Poverty in the 1990s*. Washington, DC: The Brookings Institution.

Jargowsky, Paul A., Robert Crutchfield and Scott A. Desmond, 2005. Suburban sprawl, race, and juvenile justice. In Darnell Hawkins and Kimberly Leonard, Eds., *Our Children, Their Children: Confronting Race and Ethnic Differences in American Criminal Justice*. Chicago, IL: The University of Chicago Press, pp. 167–201.

Jarvis, F. D., 1993. *Site Planning and Community Design for Great Neighborhoods*. Washington, DC: Home Builder Press.

Jeffrey, C. R., 1971. *Crime Prevention through Environmental Design*. Beverly Hills, CA: Sage.

Johnson, Alex M., Jr., 1995. *How Race and Poverty Intersect to Prevent Integration: Destabilizing Race as a Vehicle to Integrate Neighborhoods*, 143 U. Penn. L. Rev. 1595, 1620.

Johnston, Ron, David Voas and Michael Poulsen, 2003. Measuring spatial concentration: The use of threshold profiles. *Environment and Planning B: Planning and Design* 20: 3–14.

Johnston, Ron, James Forrest and Michael Poulsen, 2001. The geography of ethnicity: Residential segregation of birthplace and language groups in Sydney, 1996. *Housing Studies* 16(5): 569–594.

Kahn, Matthew E., 2001. Does sprawl reduce the Black/White housing consumption gap? *Housing Policy Debate* 12(1): 77–86.

Kain, John F., 1968. Housing segregation, negro employment, and metropolitan decentralization. *Quarterly Journal of Economics* 82(2): 175–197.

Kaplan, R. and S. Kaplan, 1989. *Experience of Nature: A Psychological Perspective*. New York: Cambridge University Press.

Karlinsky, Sarah, 2000. Community Development Corporations and Smart Growth: Putting Policy Into Practice. Neighborhood Reinvestment Corporation and the Joint Center for Housing Studies of Harvard University.

Kasarda, John, 1995. "Industrial Restructuring and the Changing Location of Jobs." pp. 215–267 in *State of the Union: America in the 1990s*, Vol. 1, edited by Reynolds Farley. New York: Russell Sage Foundation.

Katz, Bruce, 2004. *Neighborhoods of Choice and Connection: The Evolution of American Neighborhood Policy and What it Means for the United Kingdom.* Washington, DC: The Brookings Institution.

Kayden, Jerold S., 2005. Diversity by law: On inclusionary zoning and housing. In William S. Saunders, Ed., *Sprawl and Suburbia.* Minneapolis: University of Minnesota Press, pp. 71–73.

Keating, Ann Durkin, 2005. *Chicagoland: City and Suburbs in the Railroad Age.* Chicago: University of Chicago Press.

Keating, Ann Durkin, 1988. *Building Chicago: Suburban Developers and the Creation of a Divided Metropolis.* Columbus: Ohio State University Press.

Keating, W. Dennis, 1994. *The Suburban Racial Dilemma: Housing and Neighborhoods.* Philadelphia: Temple University Press.

Keating, W. Dennis, 2001. *Atlanta: Race, Class and Urban Expansion.* Philadelphia: Temple University Press.

Keating, W. Dennis, Norman Krumholz and Philip Star, 1996. *Revitalizing Urban Neighborhoods.* Lawrence, KS: University Press of Kansas.

Keels, Micere, Greg J. Duncan, Stefanie Deluca, Ruby Mendenhall and James Rosenbaum. Fifteen Years Later: Can Residential Mobility Programs Provide a Long-Term Escape From Neighborhood Segregation, Crime, and Poverty? *Demography* 42(1): 51–73.

Kefalas, Maria, 2003. *Working-Class Heroes: Protecting Home, Community, and Nation in a Chicago Neighborhood.* Berkeley: University of California Press.

Kelbaugh, Douglas S., 2002. *Repairing the American Metropolis.* Seattle: University of Washington Press, p. 287.

Kennedy, Maureen and Paul Leonard, 2001. *Dealing With Neighborhood Change: A Primer on Gentrification and Policy Choices.* Washington, DC: Brookings Institution and Policy Link.

Khadduri, Jill and Marge Martin, 1997. Mixed-income housing in the HUD multifamily housing stock. *Cityscape: A Journal of Policy Development and Research* 3(2): 33–69.

Kingsley, G. Thomas and Kathryn L. S. Pettit, 2003. Concentrated poverty: A change in course. *Neighborhood Change in Urban America* 2, May. Washington, DC: Urban Institute.

Klarman, Michael J., 2004. *From Jim Crow to Civil Rights: The Supreme Court and the Struggle for Racial Equality.* Oxford University Press.

Knox, Paul L., 1991. The Restless Urban Landscape: Economic and Sociocultural Change and the Transformation of Metropolitan Washington, DC. *Annals of the Association of American Geographers* 81(2): 181–209, 203.

Kolb, David, 2000. The Age of the List. In Gregers Algreen-Ussing, et al., Eds., *Urban Space and Urban Conservation as an Aesthetic Problem*. Rome: L'Erma di Bretschneider, pp. 27–35.

Koolhaas, Rem, 1996. *Generic City, SMLXL*. New York: The Monacelli Press.

Kostof, Spiro, 1991. *The City Shaped*. London: Thames & Hudson, Ltd.

Kraus, Neil, 2000. *Race, Neighborhoods, and Community Power: Buffalo Politics, 1934–1997*. Albany: State University of New York Press.

Krugman, Paul, 2007. *The Conscience of a Liberal*. New York: W.W. Norton.

Kunstler, James Howard, 2003. Cities of the future in the long emergency. In Emilie Buchwald, Ed., *Toward the Livable City*. Minneapolis, MN: Milkweed Editions, pp. 265–276.

Kuper, L., 1953. *Living in Towns*. London: Cresset Press.

Kupfer, Joseph, 1990. Architecture of disengagement. *Technology in Society* 12(3): 319–332.

Kushner, James A., 1982. *Apartheid in America: an Historical and Legal Analysis of Contemporary Racial Residential Segregation in the United States*. Millwood, NY: Associated Faculty Press.

Lambert, W. E. and D. M. Taylor, 1990. *Coping with Cultural and Racial Diversity in Urban America*. Westport, CT: Praeger/Greenwood.

Lamore, Rex, Terry Link and Twyla Blackmond, 2006. Renewing People and Places: Institutional Investment Policies that Enhance Social Capital and Improve the Built Environment of Distressed Communities. *Journal of Urban Affairs* 28(5): 429–442.

Lanfer, Ashley Graves and Madeleine Taylor, 2006. *Immigrant Engagement in Public Open Space: Strategies for the New Boston*. Boston: Barr Foundation. www.barrfoundation.org.

Lang, J., 2005. *Urban Design: A Typology of Procedure and Products*. Oxford, UK: Architectural Press.

Lang, R. E., J. W. Hughes and K. A. Danielsen, 1997. Targeting the suburban urbanites: marketing central-city housing. *Housing Policy Debate* 8: 437–470.

Lang, Robert E., 2004. Valuing the suburbs: Why some improvements lower home prices. *Opolis: an International Journal of Suburban and Metropolitan Studies*, 1(1), Article 3. http://repositories.cdlib.org/cssd/opolis/vol1/iss1/art3/.

Langdon, Philip, 1994. *A Better Place to Live: Reshaping the American Suburb*. Amherst: University of Massachusetts Press.

Larco, Nico, 2003. What is Urban? *Places, a Forum of Environmental Design* 15(2): 42–47.

Lazear, Edward, 2000. Diversity and Immigration. In George Borjas, Ed., *NBER Conference Volume: Issues in the Economics of Immigration*. Chicago: University of Chicago Press, 2000.

Ledebur, L. C. and W. R. Barnes, 1993. *All in it Together: Cities, Suburbs and Local Economic Regions*. Washington, DC: National League of Cities.

Lee, B. and P. Wood, 1990. The fate of residential integration in American cities: Evidence from racially mixed neighborhoods, 1970–1980. *Journal of Urban Affairs* 12(4): 425–436.

Lee, B. and P. Wood, 1991. Is neighborhood succession place specific? *Demography* 28(1): 21–40.

Lees, Loretta, 2003a. Super-gentrification: the case of Brooklyn Heights, New York City. *Urban Studies* 40: 2487–2509.

Lees, Loretta, 2003b. The ambivalence of diversity and the politics of urban renaissance: the case of youth in downtown Portland, Maine. *International Journal of Urban and Regional Research* 27(3): 613–634.

Lefebvre, Henri, 1991. *The Production of Space*. Oxford, UK: Blackwell Publishers.

LeGates, Richard and Frederic Stout, Eds., 1998. Editor's Introduction. *Early Urban Planning, 1870–1940*. New York: Routledge, p. xxxi.

Leigh, Nancy Green and Sugie Lee, 2004. Philadelphia's space in between: Inner-ring suburb evolution. *Opolis: An International Journal of Suburban and Metropolitan Studies* 1(1): 13–30.

Lennertz, Bill and Aarin Lutzenhiser, 2006. *The Charrette Handbook*. Washington, DC: APA Planners Press.

Leung, H. L., 1995. A new kind of sprawl. *Plan Canada* 35(5): 4–5.

Levy, Leo and Allen Herzog, 1978. Effects of crowding on health and social adaptation in the City of Chicago. *Urban Ecology* 3: 327–354.

Lewis Mumford Center, 2001. *Ethnic Diversity Grows, Neighborhood Integration is at a Standstill*. Report by the Lewis Mumford Center, State University of New York-Albany, December 18, 2001. http://www.mumford1.dyndns.org/cen2000/WholePop/WPreport/MumfordReport.pdf

Lipsitz, George, 2004. *Locked on This Earth: Movement and Stasis in Black Culture*. Paper presented at the symposium Constructing Race: The Built Environment, Minoritization, and Racism in the United States. University of Illinois, Urbana-Champaign, March 5–6.

Livingstone, David N., Margaret C. Keane and Frederick W. Boal, 1998. Space for religion: a Belfast case study. *Political Geography* 17(2): 145–170.

Logan, John and Harvey Molotch, 1987. *Urban Fortunes: The Political Economy of Place*. Berkeley: University of California Press.

Logan, John R., 2003. Ethnic diversity grows, neighborhood integration lags. In Bruce Katz and Robert Lang, Eds., *Redefining Urban and Suburban American: Evidence from Census 2000*. Washington: Brookings Institution Press, pp. 235–255.

Logan, John R., Brian J. Stults and Reynolds Farley, 2004. Segregation of minorities in the metropolis: Two decades of change. *Demography* 41(1): 1–22.

Low, Setha M., 2003. The edge and the center: Gated communities and the discourse of urban fear. In Setha M. Low and Denise Lawrence-Zuniga, Eds., *The Anthropology of Space and Place: Locating Culture*. Oxford: Blackwell, pp. 387–407.

Low, Setha M. and Denise Lawrence-Zuniga, 2003. Locating culture. In Setha M. Low and Denise Lawrence-Zuniga, Eds., *The Anthropology of Space and Place: Locating Culture*. Oxford: Blackwell, pp. 1–47.

Lynch, Kevin, 1960. *The Image of the City*. Cambridge, MA: Harvard University Press.

Lynch, Kevin, 1972. *What Time is this Place?* Cambridge, MA: MIT Press.

Lynch, Kevin, 1981. *Good City Form.* Cambridge, MA: MIT Press.

MacKaye, Benton, 1928. *The New Exploration: A Philosophy of Regional Planning.* New York: Harcourt, Brace and Co.

Magurran, A. E., 1988. *Ecological Diversity and its Measurement.* Princeton, NJ: Princeton University Press.

Maignan, Carole, Gianmarco Ottaviano, Dino Pinelli, Eds., 2003a. Economic growth, innovation, cultural diversity. Fondazione Eni Enrico Mattei working paper series. Available online at: http://www.feem.it/

Maignan, Carole, Gianmarco Ottaviano, Dino Pinelli and Francesco Rullani, 2003b. *Bio-Ecological Diversity vs. Socio-Economic Diversity: A Comparison of Existing Measures.* Fondazione Eni Enrico Mattei working paper series. Available online at: http://www.feem.it/

Malizia, Emil E. and Shanzi Ke, 1993. The influence of economic diversity on unemployment and stability. *Journal of Regional Science* 33(2): 221–236.

Maly, Michael T., 2000. The neighborhood diversity index: A complementary measure of racial residential settlement. *Journal of Urban Affairs* 22(1): 37–47.

Maly, Michael T., 2005. *Beyond Segregation: Multiracial and Multiethnic Neighborhoods in the United States.* Philadelphia: Temple University Press.

Maly, Michael T. and Michael Leachman, 1998. Chapter 7: Rogers Park, Edgewater, Uptown, and Chicago Lawn, Chicago. *Citiscape: A Journal of Policy Development and Research* 4(2): 131–160.

Mann, P., 1958. The socially balanced neighborhood. *Town Planning Review* 29: 91–98.

Marcuse, Peter and Ronald van Kempen, Eds., 2002. *Of States and Cities: The Partitioning of Urban Space.* Oxford, UK: Oxford University Press.

Marcuse, Peter, 2001. *Enclaves yes, Ghettoes, no: Segregation and the State.* Paper presented at the International Seminar on Segregation in the City, July 26–28, 2001, at the Lincoln Institute of Land Policy.

Marshall, Alex, 2000. *How Cities Work: Suburbs, Sprawl, and the Roads not Taken.* Austin, TX: University of Texas Press.

Marshall, Melissa J., 2004. Citizen participation and the neighborhood context: A new look at the coproduction of local public goods. *Political Research Quarterly* 57(2): 231–244.

Marshall, Stephen, 2004. *Streets and Patterns: The Structure of Urban Geometry.* London: Spon E & FN.

Martin, Judith A. and Paula R. Pentel, 2002. What the neighbors want: The Neighborhood Revitalization Program's first decade. *Journal of the American Planning Association* 68(4): 435–449.

Massey, Douglas and Mitchell Eggers, 1993. The spatial concentration of affluence and poverty during the 1970s. *Urban Affairs Quarterly* 29: 299–315.

Massey, Douglas S. and Mary J. Fischer, 2003. The Geography of Inequality in the United States, 1950–2000. In William G. Gale and Janet Rothernberg Pack, Eds., *Brookings-Wharton Papers on Urban Affairs 2003.* Washington, DC: The Brookings Institution, pp. 1–40.

Massey, Douglas S. and Nancy A. Denton, 1986. Hypersegregation in US Metropolitan Areas: Black and Hispanic Segregation Along Five Dimensions. *Demography* 26(3)(August): 373–391.

Massey, Douglas S. and Nancy A. Denton, 1993. *American Apartheid: Segregation and the Making of the Underclass.* Cambridge: Harvard University Press.

Mayer, Susan and Christopher Jencks, 1989. Growing up in poor neighborhoods: How much does it matter? *Science* 243(4897): 1441–1445.

McCarron, John, 2004. The power of sticky dots. *Planning Magazine* July: 10–13.

McConville, Shannon and Paul Ong, 2003. *The Trajectory of Poor Neighborhoods in Southern California, 1970–2000.* Washington, DC: The Brookings Institution.

McDonald, John F., 2004. The deconcentration of poverty in Chicago: 1990–2000. *Urban Studies* 41(11): 2119–2137.

McElfish, James M., Jr., 2004. *Nature-Friendly Ordinances.* Environmental Law Institute.

McHarg, Ian, 1969. *Design With Nature.* Garden City, NY: Natural History Press.

McKenzie, R.D., 1925. The ecological approach to the study of the human community. In Robert E. Park and Ernest W. Burgess, Eds., *The City: Suggestions for Investigation of Human Behavior in the Urban Environment.* Chicago: University of Chicago Press.

Mehta, Suketu, 2003. The meltingest pot. *The New York Times,* October 5, Section 6, p. 88, Column 1.

Mennel, Timothy, 2004. Victor Gruen and the construction of cold war utopias. *Journal of Planning History* 3(2): 116–150.

Metzger, John T., 2000. Planned abandonment: The neighborhood life cycle theory and national urban policy. *Housing Policy Debate* 11(1): 7–40.

Meyer, Stephen Grant, 2000. *As Long As They Don't Move Next Door: Segregation and Racial Conflict in American Neighborhood.* Oxford, UK: Rowman & Littlefield.

Meyerson, Deborah L., 2001. *Sustaining Urban Mixed-Income Communities: The Role of Community Facilities.* Urban Land Institute Land Use Policy Forum Report. Paper presented at the ULI/Charles H. Shaw Annual Forum on Urban Community Issues, October 18–19.

Meyerson, Martin and Edward C. Banfield, 1955. *Politics, Planning, and the Public Interest: The Case of Public Housing in Chicago.* Glencoe, IL: Free Press.

Michaelson, W., 1977. *Environmental Choice, Human Behavior and Residential Satisfaction.* Oxford: Oxford University Press.

Mitchell, Don, 1995. The end of public space? People's park, definitions of the public, and democracy. *Annals of the Association of American Geographers* 85(1): 108–133.

Molinaro, J., 1993. Agree on how to disagree or how to have useful discussions. *Planning Commissioners Journal* 12. http://www.plannersweb.com

Montgomery, John, 1998. Making a city: urbanity, vitality and urban design. *Journal of Urban Design* 3(1): 93–116, 98.

Morenoff, Jeffrey, Robert Sampson, and Stephen Raudenbush, 2001. Neighborhood inequality, collective efficacy, and the spatial dynamics of urban violence. *Criminology* 39: 517–560.

Morrish, William R. and Catherine R. Brown, 2000. *Planning to Stay*. Minneapolis: Milkweed Editions.

Moudon, Anne Vernez and Paul Mitchell Hess, 2000. Suburban clusters: The nucleation of multifamily housing in suburban areas of the Central Puget Sound. *Journal of the American Planning Association* 66(3): 243–264.

Moudon, Anne Vernez, 2005. Active living research and the urban design, planning, and transportation disciplines. *American Journal of Preventive Medicine* 28(2): 214–215.

Mumford, Eric, 2000. *The CIAM Discourse on Urbanism, 1928–1960*. Cambridge: MIT Press.

Mumford, Lewis, 1925. Regions – To Live In. *Survey Graphic* 7(May): 151–152.

Mumford, Lewis, 1937. What is a City? *Architectural Record*, LXXXII(November). Reprinted in Richard T. LeGates and Frederic Stout, Eds., *The City Reader*, 2nd edn. London: Routledge, pp. 93–96.

Mumford, Lewis, 1938. *The Culture of Cities*. London: Secker & Warburg.

Mumford, Lewis, 1949. *Planning for the phases of life. In The Urban Prospect*. New York: Harcourt Brace Jovanovich.

Mumford, Lewis, 1968. *The Urban Prospect*. New York: Harcourt Brace Jovanovich.

Mumford, Lewis, 1968. *Megalopolis as Anti-City, in The Urban Prospect*. New York: Harcourt Brace Jovanovich, p. 128.

Myers, Dowell, 2007. *Immigrants and Boomers: Forging a New Social Contract for the Future of America*. New York: Russell Sage Foundation.

Myerson, Deborah L., 2001. *Sustaining Urban Mixed-Income Communities: the Role of Community Facilities*. A Land Use Policy Report prepared for The Urban Land Institute, Charles H. Shaw Annual Forum on Urban Community Issues, Chicago, October 18–19.

National Research Council, 1999. *Our Common Journey: A Transition Toward Sustainability*. Washington, DC: National Academy Press.

Naveh, Z., 2004. The importance of multifunctional, self-organizing biosphere landscapes for the future of our Total Human Ecosystem – a new paradigm for transdisciplinary landscape ecology. In J. Brandt and H. Vejre, Eds., *Multifunctional Landscapes Volume I: Theory, Values and History*. Boston, MA: WIT Press, pp. 33–62.

Nechyba, Thomas, 2004. *School Finance, Spatial Segregation and the Nature of Communities: Lessons for Developing Countries?* Paper presented at the International Seminar on Segregation in the City, July 26–28, 2001, at the Lincoln Institute of Land Policy.

Nehring, Klaus and Clemens Puppe, 2002. A theory of diversity. *Econometrica* 70(3): 1155–1199.

Nelson, A. C., 1993. Disamenity influences of edge cities on exurban land values: A theory with empirical evidence and policy implications. *Urban Studies* 30: 1683–1690.

Newman, Peter W. G. and Jeffrey R. Kenworthy, 1996. The land use–transport connection: An overview. *Land Use Policy* 13(1): 1–22.

Newman, Peter W. G. and Jeffrey R. Kenworthy, 2006. Urban design to reduce automobile dependence. *Opolis: An International Journal of Suburban and Metropolitcan Studies* 2(1): 35–52.

Norberg-Schulz, Christian, 1984. *Genius Loci: Towards a Phenomenology of Architecture*. New York: Rizzoli.

Nunn, Samuel, 2001. Planning for inner-city retail development: the case of Indianapolis. *Journal of the American Planning Association* 67(2): 159–172.

Nyden, Philip, John Lukehart, Michael T. Maly and William Peterman, 1998. Chapter 1: Neighborhood Racial and Ethnic Diversity in US Cities. *Cityscape: A Journal of Policy Development and Research* 4(2): 1–17.

Nyden, Philip, Michael Maly and John Lukehart, 1997. The emergence of stable racially and ethnically diverse urban communities: A case study of nine US cities. *Housing Policy Debate* 8(2): 491–533.

Nyden, Philip, Emily Edlynn and Julie Davis, 2006. *The Differential Impact of Gentrification on Communities in Chicago*. Chicago: Loyola University Chicago Center for Urban Research and Learning.

O'Hare, William P., 1992. America's Minorities: The Demographics of Diversity. *Population Bulletin* 47(4).

Odell, Eric A., David M. Theobald and Richard L. Knight, 2003. Incorporating ecology into land use planning: the songbirds' case for clustered development. *Journal of the American Planning Association* 69(1): 72–83.

Ogletree, Charles J., Jr., 2004. *All Deliberate Speed: Reflections on the First Half Century of Brown v. Board of Education*. New York: W.W. Norton & Co.

Oliver, J. Eric, 2001. *Democracy in Suburbia*. Princeton, NJ: Princeton University Press.

Olsen, Donald J., 1986. *The City as a Work of Art*. London, Paris, Vienna. New Haven: Yale University Press.

Orfield, Myron, 2002. *American Metropolitics: The New Suburban Reality*. Washington, DC: Brookings Institution Press.

Ottensmann, J., 1995. Requiem for the tipping-point hypothesis. *Journal of Planning Literature* 10(2): 131–141.

Ottensmann, J. and M. Gleason, 1992. The movement of whites and blacks into racially mixed neighborhoods: Chicago, 1960–1980. *Social Science Quarterly* 73(3): 645–662.

Panerai, Philippe, Jean Castex, Jean Charles Depaule and Ivor Samuels, 2004. *Urban Forms: The Death and Life of the Urban Block*. Oxford: Architectural Press.

Park, Robert E., Ernest W. Burgess and Roderick D. McKenzie, 1925. *The City*. Chicago: University of Chicago Press.

Park, Robert, 1926. The Urban Community as a Special Pattern and a Moral Order. In Ernest E. Burgess, Ed., *The Urban Community*. Chicago: University of Chicago Press.

Park, Robert, 1952. *Human Communities: The City and Human Ecology*. Glencoe, IL: Free Press.

Patterson, O., 1997. *The Ordeal of Integration: Progress and Resentment in America's 'Racial' Crisis*. Washington, DC: Civitas/Counterpoint.

Pattillo-McCoy, Mary, 1999. *Black Picket Fences: Privilege and Peril Among the Black Middle Class.* Chicago: University of Chicago Press.

Peach, Ceri, 2001. *The Ghetto and the Ethnic Enclave.* Paper presented at the International Seminar on Segregation in the City, July 26–28, 2001, at the Lincoln Institute of Land Policy.

Pendall, Rolf and John I. Caruthers, 2003. Does density exacerbate income segregation? Evidence from US metropolitan areas, 1980 to 2000. *Housing Policy Debate* 14(4): 541–589.

Pendall, Rolf, 1999. *Do land-use controls cause sprawl? Environment and Planning B: Planning and Design* 26(4): 555–571.

Pendall, Rolf, 2000. Local land use regulation and the chain of exclusion. *Journal of the American Planning Association* 66(2): 125–142.

Pendall, Rolf, 2001. *Exploring connections between density, sprawl and segregation by race and income in the US metropolitan areas, 1980–1990.* Paper presented at the International Seminar on Segregation in the City, July 26–28, 2001, at the Lincoln Institute of Land Policy.

Peterman, William, 1999. *Neighborhood Planning and Community-Based Development: The Potential and Limits of Grassroots Action.* Thousand Oaks, CA: Sage Publications.

Peterman, William and Philip Nyden, 2001. Creating stable racially and ethnically diverse communities in the United States: A model for the future. *Social Policy and Administration* 35(1): 32–47.

Peterson, Ruth D., Lauren J. Krivo and Mark A. Harris, 2000. *Disadvantage and neighborhood violent crime: Do local institutions matter? Journal of Research in Crime and Delinquency* 37: 31–63.

Pickett, S. T. A., M. L. Cadenasso and J. M. Grove, 2004. Resilient cities: meaning, models, and metaphot for integrating the ecological, socio-economy and planning realm. *Landscape and Urban Planning* 69: 369–384.

Pielou, E. C., 1975. *Ecological Diversity.* New York: John Wiley & Sons.

Pinedo-Vasquez, Miguel, Jose Barletti Pasqualle, Dennis Del Castillo Torres and Kevin Coffey, 2002. A tradition of change: The dynamic relationship between biodiversity and society in sector Muyuy, Peru. *Environmental Science & Policy* 5: 43–53.

Polese, Mario and Richard Stren, Eds., 2000. *The Social Sustainability of Cities: Diversity and the Management of Change.* Toronto: University of Toronto Press.

Polyzoides, Stefanos, Roger Sherwood and James Tice, 1997. *Courtyard Housing in Los Angeles: A Typological Analysis.* Princeton: Princeton Architectural Press.

Popkin, Susan J., Diane K. Levy, Laura E. Harris, Jennifer Comey and Mary K. Cunningham, 2004. The HOPE VI Program: What about the residents? *Housing Policy Debate* 15(2): 385–414.

Powell, John A., 1999. Race, poverty, and urban sprawl: Access to opportunities through regional strategies. *Forum for Social Economics* 28(2): 1–20.

Powell, John A., 2002. Sprawl, fragmentation and the persistence of racial inequality. In Gregory D. Squires, Eds., *Urban Sprawl: Causes, Consequences and Policy Responses.* Washington, DC: Urban Institute Press, pp. 73–117.

Powell, John A., 2003. Opportunity-based housing. In Emilie Buchwald, Ed., *Toward the Livable City*. Minneapolis, MN: Milkweed Editions, pp. 181–211.

Project for Public Spaces, 2007. Great Public Spaces. http://www.pps.org/gps

Prosper, Vera, 2004. Aging in place in multifamily housing. *Cityscape: A Journal of Policy Development and Research* 7, 1. http://www.huduser.org/periodicals/cityscpe/vol7num1/index.html

Puentes, Robert and David Warren, 2006. *One-Fifth of America: A Comprehensive Guide to America's First Suburbs*. Washington, DC: The Brookings Institution.

Punter, J., 1991. Participation in the design of urban space. *Landscape Design* 200: 24–27.

Putnam, R. D., 2000. *Bowling Alone. The Collapse and Revival of American Community*. New York: Simon and Schuster.

Putnam, Robert and Lewis M. Feldstein, 2003. *Better Together: Restoring the American Community*. New York: Simon & Schuster.

Pyatok, Michael, 2002. The Narrow Base of the New Urbanists. Progressive Planning Magazine. Spring Issue. http://www.plannersnetwork.org/publications/mag_2002_2_spring.html

Pyatok, Michael, 2005. Designing for affordability: An architect's perspective. In Adrienne Schmitz, Ed., *Affordable Housing: Designing an American Asset*. Washington, DC: Urban Land Institute, pp. 31–35.

Qadeer, Mohammad A., 1997. Pluralistic planning for multicultural cities: the Canadian practice. *Journal of the American Planning Association* 63(4): 481–494.

Quercia, Roberto G. and George Galster, 1997. Threshold effects and the expected benefits of attracting middle-income households to the central city. *Housing Policy Debate* 8(2): 409–435.

Quigley, J., 1998. Urban diversity and economic growth. *Journal of Economic Perspectives* 12(2): 127–138.

Quigley, John M. and Steven Raphael, 2003. Is housing affordable? Why isn't it more affordable? *Journal of Economic Perspectives* 16(4).

Quinn, Lois M. and John Pawasarat, 2003. *Racial integration in urban America: A block level analysis of African American and White housing patterns*. Employment and Training Institute, University of Wisconsin-Milwaukee.

Rapoport, Amos, 1977. *Human Aspects of Urban Form*. Oxford: Pergamon.

Rapoport, Amos, 1990. *The Meaning of the Built Environment. A Nonverbal Communication Approach*. Tucson, AZ: University of Arizona Press.

Real Estate Research Corporation, 1982. *Urban Infill: Its Potential as a Development Strategy*. Washington: US Department of Housing and Urban Development.

Reardon, Sean F. and Glenn Firebaugh, 2002. Response: Segregation and social distance – a generalized approach to segregation measurement. *Sociological Methodology* 32: 85–101.

Reeves, Dory, 2005. *Planning for Diversity: Policy and Planning in a World of Difference*. London: Routledge.

Regen, K., 1990. Aesthetic zoning. *Fordham Law Review* 58: 1013–1031.

Rehn, Steven D., 2002. *Measuring Segregation in Melting-Pot Suburbs: Concepts, Methods and a Case Study.* Unpublished thesis, Department of Planning, Public Policy & Management, University of Oregon.

Reissman, Leonard, 1964. *The Urban Process: Cities in Industrial Societies.* New York: Free Press of Glencoe.

Richmond LISC, 2005. *The Ripple Effect: Economic Impacts of Targeted Community Investments.* Richmond, VA: Richmond LISC. http://www.lisc.org/content/publications/detail/762/

Rishbeth, C., 2001. Ethnic minority groups and the design of public open space: An inclusive landscape? *Landscape Research* 26(4): 351–366.

Ritzdorf, Marsha, 1997. Family values, municipal zoning, and African American family life. In June Thomas and Marsha Ritzdorf, Eds., *Urban Planning and the African American Community: In the Shadows.* Thousand Oaks, CA: Sage Publications, pp. 75–89.

Robertson, Kent A., 1997. Downtown retail revitalization: a review of American development strategies. *Planning Perspectives* 12: 383–401.

Robinson, Charles Mulford, 1916. *City Planning.* New York: G.P. Putnam's Sons.

Rohe, William and Lance Freeman, 2001. Assisted housing and residential segregation. *Journal of the American Planning Association* 67: 279–292.

Romaya, Sam and Carole Rakodi, Eds., 2002. *Building Sustainable Urban Settlements: Approaches and Case Studies in the Developing World.* London: ITDG Publishing.

Romaya, Sam and Carole Rakodi, Eds., 2003. *Building Sustainable Urban Settlements: Approaches and Case Studies in the Developing World.* London: ITDG Publishing.

Rose, Dina R., 2000. Social disorganization and parochial control: Religious institutions and their communities. *Sociological Forum* 15: 339–358.

Rosenbaum, James E., Linda K. Stroh and Cathy A. Flynn, 1998. Lake Parc Place: A study of mixed-income housing. *Housing Policy Debate* 9(4): 703–740.

Rothschild, Joan and Alethea Cheng, 1999. *Design and Feminism – Re-Visioning Spaces, Places, and Everyday Things.* New Brunswick: Rutgers University Press.

Rowe, Colin and Frederick Koetter, 1978. *Collage City.* Cambridge: MIT Press.

Rowley, Alan, 1996. Mixed-use development: Ambiguous concept, simplistic analysis and wishful thinking? *Planning Practice and Research* 11(1): 85–97.

Ruddick, Susan, 1996. Constructing difference in public spaces: race, class and gender as interlocking systems. *Urban Geography* 17(2): 132–151.

Ryan, William, Allan Sloan, Mania Seferi and Elaine Werby, 1974. *All in it Together: An Evaluation of Mixed-Income Multi-Family Housing.* Boston: Massachusetts Housing Finance Agency.

Rybczynski, Witold, 2000. Where Have all the Planners Gone? In Lloyd Rodwin and Bish Sanyal, Eds., *The Profession of City Planning: Changes, Successes, Failures and Challenges (1950–2000).* Rutgers, NJ: Center for Urban Policy Research, 1999, pp. 210–216.

Saarinen, Eliel, 1943. *The City: Its Growth, Its Decay, Its Future.* New York: Reinhold Publishing Co.

Sailer, Steve, 2007. Fragmented Future: Multiculturalism doesn't make vibrant communities but defensive ones. *The American Conservative* January 15.

Salingaros, Nikos A., 1998. Theory of the urban web. *Journal of Urban Design* 3: 53–71.

Saltman, Juliet, 1990. *A Fragile Movement*. New York: Greenwood.

Sampson, Robert J. and Steve Raudenbush, 1999. Systematic Social Observation of Public Spaces: A New Look at Disorder in Urban Neighborhoods. *American Journal of Sociology* 105: 603–651.

Sampson, Robert J., Jeffrey D. Morenoff and Felton Earls, 1999. Beyond social capital: Spatial dynamics of collective efficacy for children. *American Sociological Review* 64(5): 633–660.

Sampson, Robert J., Stephen Raudenbush and Felton Earls, 1997. Neighborhoods and Violent Crime: A Multilevel Study of Collective Efficacy. *Science* 277: 918–924.

Sandercock, Leonie, 1998. *Towards Cosmopolis*. Chichester, UK: John Wiley & Sons.

Sanders, Marion, 1970. *The Professional Radical: Conversations with Saul Alinsky*. New York: Harper and Row.

Sang, Yan and Gerrit-Jan Knaap, 2004. Measuring urban form: Is Portland winning the war on sprawl? *Journal of the American Planning Association* 70(2): 210–225.

Sarkissian, S., 1976. The idea of social mix in town planning: An historical overview. *Urban Studies* 13(3): 231–246.

Sassen, Saskia, 1997. *Cities: Between Global Actors and Local Conditions*. LeFrak monograph. College Park, MD: Urban Studies and Planning Program, University of Maryland.

Saxenian, A., 1999. *Silicon Valley's new immigrant entrepreneurs*. Berkeley: Public Policy Institute of California.

Scheer, Brenda Case, 2001. The Anatomy of Sprawl. *Places* 14(2): 28–37.

Schelling, Thomas, 1978. *Micromotives and Macrobehavior*. New York: W.W. Norton.

Schill, Michael H. and Susan Wachter, 1995. The Spatial Bias of Federal Housing Law and Policy: Concentrated Poverty in Urban America. *University of Pennsylvania Law Review* 143(5).

Schneider, Mark and Thomas Phelan, 1993. Black suburbanization in the 1980s. *Demography* 30(2): 269–279.

Schubert, Michael F. and Alison Thresher. Lessons from the field: three case studies of mixed-income housing development. University of Illinois at Chicago: Great Cities Institute. On-line at http://www.uic.edu/cuppa/gci/.

Schwartz, Alex and Kian Tajbakhsh, 1997. Mixed-income housing: Unanswered questions. *Cityscape: A Journal of Policy Development and Research* 3(2): 71–92.

Scott, James C., 1998. *Seeing Like a State: How Certain Schemes to Improve the Human Condition Have Failed*. New Haven, CT: Yale University Press.

Sennett, Richard, 1970. *The Uses of Disorder: Personal Identity and City Life*. New York: Alfred A. Knopf.

Sennett, Richard, 1994. *Flesh and Stone: The Body and the City in Western Civilization*. New York: W.W. Norton.

Sharfstein, D. and J. Stein, 2000. The dark side of internal capital markets: divisional rent-seeking and inefficient investment. *Journal of Finance* 55: 2537–2564.

Shen, Qing, 2001. A spatial analysis of job openings and access in a US metro-politan area. *Journal of the American Planning Association* 67(1): 53–68.

Silver, Christopher, 1985. Neighborhood planning in historical perspective. *Journal of the American Planning Association* 51(2): 161–174.

Simpson, E., 1949. Measurement of diversity. *Nature* 163: 688.

Singer, Audrey, 2004. *The Rise of New Immigrant Gateways*. Washington, DC: The Brookings Institution.

Sirianni, Carmen and Lewis Friedland, 2001. *Civic Innovation in America: Community Empowerment, Public Policy, and the Movement for Civic Renewal*. Berkeley, CA: University of California Press.

Skerry, Peter, 2002. Beyond sushiology: does diversity work? *The Brookings Review* 20(1): 20–24.

Skjaeveland, Oddvar and Tommy Garling, 1997. Effects of interactional space on neighbouring. *Journal of Environmental Psychology* 17: 181–198.

Slater, Tom, Winifred Curran and Loretta Lees, 2004. *Environment and Planning A* 36: 1141–1150.

Slater, Tom, 2004. North American gentrification? Revanchist and emancipatory perspectives explored. *Environment and Planning A* 36: 1191–1213.

Smith, Alastair 2002. *Mixed-Income Housing Developments: Promise and Reality*. Neighborhood Reinvestment Corporation: Joint Center for Housing Studies of Harvard University. On-line at http://www.jchs.harvard.edu/publications/W02-10_Smith.pdf

Smith, Janet L., 2001. Mixing It Up: Public Housing Redevelopment in Chicago. Paper presented at the Conference Area-Based Initiatives in Contemporary Urban Policy, Copenhagen, May 17–19, 2001. On-line at http://www.by-og-byg.dk/eura/workshops/papers/workshop1/smith.pdf

Smith, Neil and Peter Williams, Eds., 1986. *Gentrification of the City*. London: Allen and Unwin.

Smith, Neil, 1996. *The New Urban Frontier: Gentrification and the Revanchist City*. London: Routledge.

Smith, Richard A., 1993. Creating stable racially integrated communities. *Journal of Urban Affairs* 15(2): 115–140.

Smith, Richard A., 1998. Discovering stable racial integration. *Journal of Urban Affairs* 20(1): 1–25.

Smith, Tara, Maurice Nelischer and Nathan Perkins, 1997. *Landscape and Urban Planning* 39: 229–241.

Soja, Edward, 1989. *Postmodern Geographies: The Reassertion of Space in Critical Social Theory*. London: Verso.

Soja, Edward, 2000. *Postmetropolis: Critical Studies of Cities and Regions*. Oxford: Blackwell.

Sorkin, Michael, 1999. Introduction: Traffic in Democracy. In Joan Copjec and Michael Sorkin, Eds., *Giving Ground: The Politics of Propinquity*. London: Verso, pp. 1–15.

Sorkin, Michael, 2006. The end(s) of urban design. *Harvard Design Magazine* 25: 5–18.

Southworth, Michael and Eran Ben-Joseph, 2003. *Streets and the Shaping of Towns and Cities.* Washington, DC: Island Press.

Spain, Daphne, 1993. Been-heres versus come-heres. *Journal of the American Planning Association* 59(2): 156–172.

Squires, Gregory D., Ed., 2002. *Urban Sprawl: Causes, Consequences & Policy Responses.* Washington, DC: The Urban Institute Press.

Squires, Gregory D., Samantha Friedman and Catherine E. Saidat, 2001. *Housing Segregation in the United States: Does Race Matter?* Paper presented at the International Seminar on Segregation in the City, July 26–28, 2001, at the Lincoln Institute of Land Policy.

Stamps, Arthur, 2003. Advances in visual diversity and entropy. *Environment and Planning B: Planning and Design* 30: 449–463.

Stanilov, Kiril, 2002. Postwar trends, land-cover changes, and patterns of suburban development; the case of Greater Seattle. *Environment and Planning B: Planning and Design* 29: 173–195.

Stape, Andrea L., 2004. Urban planners extol diversity in downcity's buildings. *The Providence Journal* March 8.

Steinacker, Annette, 2003. Infill development and affordable housing: Patterns from 1996 to 2000. *Urban Affairs Review* 38(4): 492–509.

Steiner, Frederick, 2002. *Human Ecology: Following Nature's Lead.* Washington, DC: Island Press.

Suttles, Gerald, 1972. *The Social Construction of Communities.* Chicago, IL: University of Chicago Press.

Swanstrom, Todd, Colleen Casey, Robert Flack and Peter Dreier, 2004. *Pulling Apart: Economic Segregation Among Suburbs and Central Cities in Major Metropolitan Areas.* Washington, DC: The Brookings Institution.

Swanstrom, Todd, 2001. Regionalism reconsidered: What we argue about when we argue about regionalism. *Journal of Urban Affairs* 23(5): 479–496.

Taeuber, Karl E. and Alma F. Taeuber, 1965. *Negroes in Cities: Residential Segregation and Neighborhood Change.* Chicago: Aldine Publishing Company.

Talen, Emily, 1999. Can sense of community be built? An assessment of the social doctrine of new urbanism. *Urban Studies* 36(8): 1361–1379.

Talen, Emily, 2005. *New Urbanism and America Planning: The Conflict of Cultures.* London: Routledge.

Talen, Emily, 2003. Neighborhoods as service providers: a methodology for evaluating pedestrian access. *Environment and Planning B: Planning and Design* 30(2): 181–200.

Talen, Emily, 2006. Neighborhood-level social diversity: insights from Chicago. *Journal of the American Planning Association* 72(4): 431–446.

Tanenhaus, Sam, 2004. Black, White and Brown. *The New York Times* May 16.

Taranova, Natalya V., 2003. *The Role of the City in Fostering Intergroup Communication in a Multicultural Environment: Saint-Petersburg's Case.* Fondazione Eni Enrico Mattei working paper series. Available online at: http://www.feem.it/

Taub, Richard D., Garth Taylor and Jan Dunham, 1984. *Paths of Neighborhood Change.* Chicago: University of Chicago Press.

Taylor, P. J. and R. E. Lang, 2004. The shock of the new: 100 concepts describing recent urban change. *Environment and Planning A* 36: 951–958.

Taylor, Peter, David Walker, Gilda Catalano & Michael Hoyler, 2002. Diversity and power in the world city network. *Cities* 19(4): 231–241.

Temkin, Kenneth and William Rohe, 1996. Neighborhood change and urban policy. *Journal of Planning Education and Research* 15: 159–170.

Theil, Henri, 1972. *Statistical Decomposition Analysis.* Amsterdam: North-Holland.

Thernstrom, S. and A. Thernstrom, 1997. *American in Black and White: One Nation, Indivisible.* New York: Simon and Schuster.

Thomas, John L., 2000. Holding the Middle Ground. In Robert Fishman, Ed., *The American Planning Tradition.* Baltimore: Johns Hopkins, pp. 33–64.

Thomas, June Manning and Marsha Ritzdorf, Eds., 1997. *Urban Planning and the African American Community: In the Shadows.* Thousand Oaks, CA: Sage Publications.

Thomas, June Manning, 1997. Coming together: unified diversity for social action. In June Thomas and Marsha Ritzdorf, Ed., *Urban Planning and the African American Community: In the Shadows.* Thousand Oaks, CA: Sage Publications, pp. 258–277.

Thompson, Catharine Ward, 2002. Urban open space in the 21st Century. *Landscape and Urban Planning* 60: 59–72.

Tienda, Martha, 1991. Poor people and poor places: deciphering neighbourhood effects on poverty outcomes. In J. Haber, Ed., *Macro-Micro Linkages in Sociology.* Newberry, CA: Sage.

Tiesdell, Steven, 2003. Integrated affordable housing within market-rate developments: the design dimension. *Environment & Planning B: Planning and Design* 31(2): 195–212.

Tobler, Waldo, 1979. Cellular geography. In S. Gale and G. Olsson, Eds., *Philosophy in Geography.* Dortrecht: Riedel, pp. 379–386.

Trancik, Roger, 1986. *Finding Lost Space.* New York: Van Nostrand Reinhold.

Triggs, H. Inigo, 1909. *Town Planning: Past, Present and Possible.* London: Methuen & Co.

Tschumi, Bernard, 1996. *Architecture and Disjunction.* Cambridge: MIT Press.

Tuan, Yi-Fu, 1981. *Space and Place: The Perspective of Experience.* Minneapolis: University of Minnesota Press.

US Housing Authority, 1938. *Planning Profitable Neighborhoods.* Washington, DC: US Government Printing Office.

US Housing Authority, 1939. *Planning the Site: Design of Low-Rent Housing Projects.* Washington, DC: US Government Printing Office.

Unwin, Raymond, 1909. *Town Planning in Practice: An Introduction to the Art of Designing Cities and Suburbs.* London: T. Fisher Unwin.

Urban Land Institute, Chicago District Council and the Campaign for Sensible Growth, 2001. *Forging Partnerships: Overcoming Community Resistance to Developing Workforce Housing.* Washington, DC: The Urban Land Institute.

Vale, Lawrence J., 1998. Comment on James E. Rosenbaum, Linda K. Stroh and Cathy A. Flynn's Lake Parc Place: A study of mixed-income housing. *Housing Policy Debate* 9(4): 749.

Vale, Lawrence J., 2000. *From the Puritans to the Projects: Public Housing and Public Neighbors.* Cambridge: Harvard University Press.

Van Kempen, Ronald, 2002. The academic formulations: Explanations for the partitioned city. In: Peter Marcuse and Ronald van Kempen, Eds., *Of State and Cities: The Partitioning of Urban Space.* Oxford: Oxford University Press, pp. 35–56.

Vandell, Kerry D., 1995. Market factors affecting spatial heterogeneity among urban neighborhoods. *Housing Policy Debate* 6: 103–139.

Velez, Maria B., 2001. The role of public social control in urban neighborhoods: A multilevel analysis of victimization risk. *Criminology* 39: 837–864.

Venkatesh, Sundhir Alladi, 2002. *American Project.* Cambridge: Harvard University Press.

Venturi, Robert, Steven Izenour and Denise Scott Brown, 1977. *Learning from Las Vegas: The Forgotten Symbolism of Architectural Form.* Cambridge, MA: MIT Press.

Vigdor, Jacob L., 2002. Does gentrification harm the poor? *Brookings-Wharton Papers on Urban Affairs.* Washington: The Brookings Institution. pp. 133–182.

Von Hoffman, Alexander, 1994. *Local Attachments: The Making of an American Urban Neighborhood, 1850 to 1920.* Baltimore: Johns Hopkins University Press.

Wallace, M. and B. Milroy, 1999. Intersecting claims: planning in Canada's cities. In T. Fenster, Ed., *Gender, Planning and Human Rights.* London: Routledge.

Walljasper, Jay, 2007. *The Great Neighborhood Book: A Do-it-Yourself Guide to Placemaking.* Gabriola Island, BC, Canada: New Society Publishers.

Wang, Vincent, 1996. Mixed use: the answer to the successful urban environment. *Built Environment* 22(4): 312–314.

Warren, Mark R., 2001. *Dry Bones Rattling: Community Building to Revitalize American Democracy.* Princeton, NJ: Princeton University Press.

Wassmer, Robert W., 2001. *The Economics of the Causes and Benefits/costs of Urban Spatial Segregation.* Paper presented at the International Seminar on Segregation in the City, July 26–28, 2001, at the Lincoln Institute of Land Policy.

Webber, M. M., 1963. Order in diversity: community without propinquity. In L. Wingo, Eds., *Cities and Space: The Future Use of Urban Land.* Baltimore: Johns Hopkins University Press, pp. 23–54.

Webber, Melvin M., 2004. Spread-city everywhere. *Access* 24(1).

Weiss, Marc, 1987. *The Rise of the Community Builders: The American Real Estate Industry and Urban Land Planning.* New York: Columbia University Press.

Weiss, Michael J., 1988. *The Clustering of America.* New York: Harper & Row.

Weiss, Michael J., 2000. *The Clustered World: How We Live, What We Buy and What it All Means About Who We Are.* Boston: Little, Brown & Co.

Welch, Susan, Lee Sigelman, Timothy Bledsoe and Michael Combs, 2001. *Race and Place: Race Relations in an American City.* Cambridge: Cambridge University Press.

Wells, Amy Stuart, Jennifer Jellison Holme, Anita Tijerina Revilla and Awo Korantemaa Atanda, 2004. *How Desegregation Changed Us: The Effects of*

Racially Mixed Schools on Students and Society. Available on-line at http://www. teacherscollege.edu/newsbureau/features/wells033004.htm

West, Cornel, 1991. Diverse New World. Democratic Left 19.4. Reprinted in *Debating PC: The Controversy over Political Correctness on College Campuses*, ed. Paul Berman. New York: Laurel/Dell, 1992. Reprinted in *Signs of Life in the USA: Readings on Popular Culture for Writers*, Sonia Maasik and Jack Solomon, Eds. Boston: Bedford Books, 1997, pp. 557–562.

White, Michael J., 1986. Segregation and diversity measures in population distribution. *Population Index* 52: 198–221.

Wiese, Andrew, 2004. *Places of Their Own: African American Suburbanization in the Twentieth Century.* Chicago: University of Chicago Press.

Williams, K. Y. and C. A. O'Reilly, 1998. Demography and diversity in organizations: A review of 40 years of research. In B. Staw and R. Sutton, Eds., *Research in Organizational Behavior*, Vol. 20. Greenwich, CT: JAI Press, pp. 77–140.

Wilson, E. O., 1988. *Biodiversity.* Washington, DC: National Academy Press.

Wilson, William Julius, 1987. *The Truly Disadvantaged: The Inner City, the Underclass and Public Policy.* Chicago: University of Chicago Press.

Wilson, William Julius, 1996. *When Work Disappears: The World of the New Urban Poor.* New York: Alfred A. Knopf.

Wilson, William Julius and Richard P. Taub, 2006. *There Goes the Neighborhood: Racial, Ethnic, and Class Tensions in Four Chicago Neighborhoods and Their Meaning for America.* New York: Alfred A. Knopf.

Wirth, Louis, 1938. Urbanism as a way of life. *American Journal of Sociology* XLIV(1).

Wong, David W. S., 2002. Spatial measures of segregation and GIS. *Urban Geography* 23: 85–92.

Wright, David J., 2001. *It Take a Neighborhood: Strategies to Prevent Urban Decline.* Albany, NY: The Rockefeller Institute Press.

Wyly, Elvin and Daniel Hammel, 2004. Gentrification, segregation, and discrimination in the American urban system. *Environment and Planning A* 36: 1215–1241.

Yeomans, Alfred B., 1916. *City Residential Land Development: Studies in Planning.* Chicago: University of Chicago Press.

Yinger, John, 1995. Opening Doors: How to Cut Discrimination by Supporting Neighborhood Integration. Syracuse University: Maxwell School of Citizenship and Public Affairs, Center for Policy Research. Policy Brief.

Young, I. M., 2000. A critique of integration as the remedy for segregation. In D. Bell and A. Haddour, Eds., *City Visions.* London: Prentice Hall.

Zachary, Pascal G., 2000. *The Global Me: New Cosmopolitans and the Competitive Edge – Picking Globalism's Winners and Losers.* New York: Perseus Books.

Zielenbach, Sean, 2000. *The Art of Revitalization: Improving Conditions in Distressed Inner-City Neighborhoods.* New York: Garland Publishing.

Zorbaugh, Harvey Warren, 1929. *The Gold Coast and the Slum: A Sociological Study of Chicago's Near North Side.* Chicago: University of Chicago Press.

Zucker, Paul, 1959. *Town and Square: From the Agora to the Village Green*. New York: Columbia University Press.

Zukin, Sharon, 1982. *Loft Living: Culture and Capital in Urban Change*. New Brunswick, NJ: Rutgers University Press.

Zukin, Sharon, 1995. *The Cultures of Cities*. Oxford: Blackwell.

Zukin, Sharon, 1998. Urban lifestyles: Diversity and standardization in spaces of consumption. *Urban Studies* 35(5): 825–839.

INDEX